The Metahistory of Western Knowledge in the Modern Era

I0660564

The Metahistory of Western Knowledge in the Modern Era

FOUR EVOLVING METAPARADIGMS, 1648 TO PRESENT

Mark E. Blum

ANTHEM PRESS

Anthem Press
An imprint of Wimbledon Publishing Company
www.anthempress.com

This edition first published in UK and USA 2025
by ANTHEM PRESS
75–76 Blackfriars Road, London SE1 8HA, UK
or PO Box 9779, London SW19 7ZG, UK
and
244 Madison Ave #116, New York, NY 10016, USA

First published in the UK and USA by Anthem Press in 2021

British Library Cataloguing-in-Publication Data
A catalogue record for this book is available from the British Library.

Library of Congress Control Number: 2024949822
A catalog record for this book has been requested.

ISBN-13: 978-1-83999-476-0 (Pbk)
ISBN-10: 1-83999-476-2 (Pbk)

This title is also available as an e-book.

CONTENTS

INTRODUCTION

One of the most interesting metahistorical investigations would be an inquiry into the great historical rhythms.[1]

Ortega y Gasset

1

What is a "metahistory"?

When one organizes the events over periods of years, and gives it an appellation such as "Modernism," the organization of facts is guided by concepts and values discerned throughout these periods, comparable facts sufficient to call it an "era," or an "epoch," or other terms that insist on the shared aspects of those years, regardless of differences seen as well over the span considered. One can call such an effort a "metahistory," in that what is tracked is not merely human events that are political, economic, ideological, sociological or other disciplinary descriptors, but an overview that critically links all the years under consideration. Even more, to have a "metahistory" is to discern how the people of eras, epochs or the other organizational labels thought. Human history is generated by choices, choices informed by intuitions and more intentional understandings. One of the aspects I will dwell upon in this "metahistory" of Modernism is the presence of "perspective," how one sees in a time what is there to be addressed and dealt with. Perspectives can be poorly informed or in their very nature not adequate for a sufficient knowledge of what is addressed, even as one must as a human judge what faces one. To discern from evidence how one's perspective configures an event is the "meta" of "metahistory." To have "meta" knowledge is this comprehension of the scope and benefits, yet limitations, of one's "perspective" and that of others.[2] Only a historian interested in such perspectives can be called a "metahistorian."

---

1. Ortega y Gasset, "Die Aufgabe unserer Zeit" (The Task of our Time), in Gesammelte Werke, trans. Helene Weyl and Ulrich Weber, 4 vols. (Stuttgart: Deutsche Verlags-Anstalt, 1950), 2: 84.
2. The coinage of the term insofar as an insight into how historical perspectives are constructed was by Ortega y Gasset. He coined the term in 1923. *The Oxford English Dictionary* does not list his contribution of the 1920s, but I include in this endnote the general definition of "being aware of the systems of critical awareness" from the *Oxford English Dictionary*, meta-system *n.* 1956.

Wilhelm Dilthey wrote of metahistorical rhythms in history earlier than Ortega y Gasset, while not using the concept explicitly. The metaphorical "historical rhythms" meant for Ortega y Gasset as well as Dilthey how humans individually and collectively perceived and acted toward what was necessary in formulating the problems and necessary actions of a time in a society, yet also the changing emphases that created a "rhythm." What is a "historical rhythm"? When Bismarck wrote that a collective emphasis in societal feeling and societal policy occurred at the end of centuries in contradistinction to when one focused upon individuals in making policy in a state, asserting that the individual emphasis occurred in the middle of a century that was a "historical rhythm" in his eyes.[3] Even though I will take issue with his location of collective action at the end of centuries and individual action in the middle, for Bismarck that made him for this instance a self-aware "metahistorian."

The "metahistorical" evidence of Dilthey and Ortega y Gasset came not only from individuals in various disciplines that were constructed to gain and act upon knowledge, but especially historians whose views were always "meta" in the sense of showing "change over time" in the values and understandings of a society. While one can point to any competent historian and see their overview of what occurred as a "meta" understanding, Dilthey and Ortega y Gasset were of a new kind of discernment—one that created some distance from the norms of historical interpretation that were conditioned by their own time in its constraints of not only general knowledge, but self-knowledge. Knowing of the conceptual orientation of not only individuals, but of the normative interpretative historiography of a time, is the kind of "metahistory" that has not yet been thoroughly executed.

My efforts in this text are aided by many who have begun such studies, such as Dilthey and Ortega y Gasset, as well as the more thorough efforts into "metahistorical rhythms" such as the nineteenth century historiography by Hayden White. In his *Metahistory: The Historical Imagination in Nineteenth-Century Europe* White opens up new interpretive tools for generating his "meta" perspective on the several periods of nineteenth century historical understandings—including the use of stylistics and insight into the structural logic of events guided by the semantics and syntactical usage of historical minds in the several periods of the nineteenth century. Both the stylistics of semantical terms that guided thought in a period of time, and the syntactical logic of the structure of events, an off-shoot of stylistics, will be central in this text. White used the syntactical logical inquiries of Stephen C. Pepper,[4] a mid-twentieth century philosopher of history, to guide his own thought.[5]

    J.H. Woodger tr. A. Tarski *Logic, Semantics, Metamath*, 116. It is possible to construct a particular science, namely the "metasystem," in which the given system is subjected to investigation. https://wwcom.echo.louisville.edu/view/Entry/117150?rskey=lxQ2Z1&result=4#eid

3. See Heinrich Friedjung, "Appendix I," in *The Struggle for Supremacy in Germany, 1859–1866*, trans. A.J.P. Taylor and W.L. McElwee (London: Macmillan & Co., 1935), p. 315.

4. Stephen C. Pepper, *World Hypotheses, Prolegomena to systematic philosophy and a complete survey of metaphysics* (Berkeley, Calif.: University of California Press, 1942; 1970).

5. Hayden White, *Metahistory: The Historical Imagination in Nineteenth-Century Europe* (Baltimore: Johns Hopkins University, 1973), 13–21. See also my use of Stephen C. Pepper in informing the

The fundamental perspective for conceiving "metahistory" can be found in largely German thinkers. The German historicism of Herder, followed by that of Kant—who took issue with Herder's "metahistorical" interpretations, and then Kant's own students who deepened the evidence for the generational history that is the foundation for "metahistory," Friedrich Schleiermacher, Georg Wilhelm Friedrich Hegel, Leopold von Ranke and finally Friedrich Meinecke. Yet these thinkers were un-self-consciously of their culture, failing to comprehend the narrative norms that locked them into a perspective generated around them, without them appreciating its full affect upon their own thinking. An accomplished "metahistory" liberates even the person within such norms to see fully his or her inherited perspective with a "meta" insight into its beginnings and its development. The literature of the nineteenth century, as well as the literature that surrounded White in the late twentieth- and early twenty-first century, enabled him to see the narrative norms of *historia*—from which he wrote and of which he wrote of in his purview of nineteenth century historians and their event-structures as well as their historiographical purview. White's emphasis on the story-telling aspect of what Herodotus called *historia*—inquiry and a human proportioned narrative of beginnings, middles and endings—gave him access to the great metahistorical rhythms of how people viewed and spoke of the structure of historical events.

Let me conclude my definition of what I intend as a "metahistory" with a brief discussion of Louis O. Mink's understanding of "configural comprehension." "Configural comprehension" is a useful concept in comprehending "metahistory," and in particular the discernment of metaparadigms, a form of "metahistory" that coheres periods of years with a constant design. Mink states that one must comprehend a span of cultural history as one might see the course of a river from above, charting in one's mind the design of the span of river perceived as a cohering a pattern that explains historical activity within that design:

> In the configurational comprehension of a story … the end is connected with the promise of the beginning as well as the beginning with the promise of the end, and the necessary backward reference cancels out, so to speak, the contingency of the forward references. To comprehend temporal succession means to think of it in both directions at once, and then time is no longer the river which bears us along but the river in aerial view, upstream and downstream in a single survey.[6]

---

logical structure of events in Mark E. Blum, *Continuity, Quantum, Continuum, and Dialectic: The Foundational Logics of Western Historical Thinking* (New York: Peter Lang, 2006), 263–308, and Mark E. Blum, *Cognition and Temporality, The Genesis of Historical Thought in Perception and Reasoning* (New York: Peter Lang, 2019), 187–201.

6. See this quotation and the discussion of it by Frank, Ankersmit, *Meaning, Truth, and Reference in Historical Representation* (Ithaca, New York: Cornell University Press, 2012), 36–40. See also F.R. Ankersmit, *Historical Representation* (Stanford, Calif.: Stanford University Press, 2001), 203–5 where Ankersmit offers the same quote from Mink as Ankersmit introduces Erich Auerbach's term "figura" that is realized in a configurational comprehension. Ankersmit discusses the "realism" captured by the historian by seeing the narrative form that carries the linked facts into a structural design. The "metaparadigm" must be conceived as such a realistic ordering of facts and their implications.

Mink's configurational comprehension enables him to compose a coherent historical narrative as a "metahistorian." The narrative is established from attention to what surfaces through an inquiry that more and more is guided by the very design that emerges gradually from the facts discerned. The facts begin to be recognized as a pattern, the pattern is seen again and again. Human practices in their intentions and modes of behavior, what becomes normative of a time, is Mink's river in its configuration. One can first discern the end, or the beginning, or even the middle. Every factual discernment reveals more and more of the configurational design, akin to Michelangelo's speaking of sculpting as revealing the figure contained in the stone. One can read Mink's trope of the river's design as a configurational comprehension that fulfills Herodotus's concept of *historia*—history as inquiry into the facts of human events that becomes a story of human intentions and their consequences which can be seen as a coherent period of years, with a beginning, a middle and an end.[7]

Mink's river or Michelangelo's figure emerging from the stone is an analogue to the "metahistory" that is this text. The configurations of this "metahistory" will be offered from the aerial view I have taken of almost four centuries—from the mid-1600s to the first decades of the 21st century—that of Modernism.

2

The metahistory of Modernism: How is it approached methodologically?

The "metahistory" I present uses evidence of the thought of individuals who lived in the Modern Era who spoke about their times as well as earlier times in their accounts of the reality around them and before them. The Modern Era can be dated in its origins from the Peace of Westphalia, negotiated from 1643 to 1648. The formal end of the religious wars became a watershed period for a major change in thinking that spanned all the arts and the sciences. There was an Early Modernism from roughly the 1570s until 1648 that established evidence into inorganic and organic nature, as well as human thought, but the final and formal causes were still primarily understood as a "natural law" occasioned by the Divine. In the Early Modern period inquirers into the natural world and human nature included observant, self-aware, penetrating thinkers such as Galileo, Montaigne, Grotius, Descartes and Pascal, who did lay an intellectual foundation for the generations of Modernism to come. What I see as their major contribution, other than the fecund

---

7. Herodotus, *The Histories*, trans. Aubrey de Sélincourt, ed. A.R. Burn (Harmondsworth, Middlesex, England: Penguin, 1972). The term used for his work was *historia*. This meant both inquiry and the narrative story of its facts. The editor, A. R. Burn, writes of this term: "Herodotus of Halicarnassus: Researches (*historia*). These words visible when the papyrus was rolled up, served the purpose of those on our book-covers. 'Research' or 'inquiry', a word often used in the text is in Greek *historia*; and its specialized meaning of *history* was born there. Herodotus goes on, in the manner of an expansive early modern title-page, to state his purpose: 'that the great deeds of men may not be forgotten … whether Greeks or foreigners; especially the causes of war between them'. (p. 7)."

conceptual orientations for problem formulation and solution in their fields of knowing, can be stated in a succinct axiom. They taught the Western mind "one must become aware of how one is aware in order to comprehend the perspectives of self and others." This I see as one of the major shared understandings in all Modernism that followed, and in which we are presently. Knowing the presence and scope of our own perspective opens up our knowing of the differing perspectives of others.

To conduct a metahistorical investigation of how individuals have written of the development knowledge in all fields of human knowing one must comprehend the causes in view with tangible evidence accessible to all. Aristotle called these causes, if they were to provide necessary and sufficient evidence, the final, formal, material and effi-cient causes. While major insights realized by these Early Modern writers would become increasingly separate disciplines in the arts and the sciences, creating a ground for further discoveries based upon what had been understood, their "early" modern insights were evidentially limited. Early Modernism limited one's insight into "efficient causes," the cause-effect sequences of material and immaterial existence that were empirically verifi-able in their evidence. Cause and effect understandings that were wholly secular in their final, formal, material and efficient causes were realized only after such post-Peace of Westphalia public policies as the separation of church and state.[8] The tangible evidence of nature and the human bracketed from the role of the Divine became the norm. The transition away from Early Modernism sought in the natural world for causal sequences that could be verified by others, establishing the scientific and rational bases of one objec-tive truth in all areas of what became the arts and the sciences. Who we are, what we are, what we do and how we do it, what surrounds us, how it affects us, when and why it affects us—by the late twentieth century—is an unquestioned progress in knowing into our present time.

Pascal's first letter in his 1656 *Provincial Letters* that castigates the misuse of the term "proximate power" by differing visions within the Catholic faith qualifies as an astute descriptive account of those who are aware of being aware and of those who are not.[9] Yet, this not the first instance of "being aware of how one is aware" that enables us to see

---

8. My use of the Aristotelian "four causes"—the final, formal, material and efficient—will be central to how I demarcate the several phases of each evolving metaparadigm. All four play a role in any problematic and its solution in the arts and the sciences. And, the prominence of one in certain phases of a metaparadigm will justify my discussion of that phasal address by thinkers. I will further develop my use of these causes below. See Aristotle, Poetics for time, place, and manner. I draw from the general discussion in my use of these categories. Fictive lives over the centuries in the poetics of the German or Austrian-German are highly relevant to my text, indeed just as historical writing or historiographical theory. That is because of the narrative forms normative of a time. As the same categorical form, the reader must extrapolate the terms of Aristotle's discussion of varieties of time, place and manner in drama and other poetics to the factual narratives are composed in any national *historia*. The four causes—the final, formal, efficient and material—are to be found in Aristotle's Physics, Book II, Chapter III, and his *Metaphysics*, Book 5, 2.

9. Blaise Pascal, *Pascal's Pensées and the Provincial Letters* (New York: The Modern Library, 1941), 325–55.

how we or others use a conceptual perspective in the Modern era comes from an even earlier Early Modern understanding of how one's conscious mind is governed by conceptual understandings that one can fail to appreciate insofar how one uses it as opposed to others. Montaigne's essay on "conversation" is one such Early Modern example,[10] and, Pascal himself in his initial two *Pensées,* presumably written early in his life, offers a clear discussion of how one is self-aware of his or her perspective, as well as of the way others use concepts and form perspectives (Pascal 1941, 3–5).

Removing the Divine as the final cause of thought and action within human history did not invalidate religion. In what I will describe as the Second Modern metaparadigm, Immanuel Kant laid out sufficient proofs that while God may exist and his hand in who we are could possibly be, secular scientific thought could not rely on such a final, formal, material or efficient cause.[11] Rather, human goal setting, which I will describe as problem formulation, had a final cause of addressing issues and problems in one's "life world" (as Edmund Husserl would so call the human and natural environ of one's every existence).[12]

---

10. Michel de Montaigne, "The Art of Conversation," in *The Complete Essays,* trans. M.A. Screech (London: Penguin Books, 1991), 1044–69.

11. My choice among Kant's various writings that touch upon this issue of an adequate secular conception of the Aristotelian four causes would one appreciate how to solve civilization's problems is the essay "Idea for a Universal History From a Cosmopolitan Point of View," in *On History,* trans. Lewis White Beck (New York: Bobbs-Merrill, 1963), 11–26. Kant's statement at the Introduction to the essay puts the final cause in human hands quite strongly as well as cynically. He writes: "Since the philosopher cannot presuppose any (conscious) individual purpose among men in their great drama, there is no other expedient for him except to try to see if he can discern a natural purpose in this idiotic course of things human. In keeping with this purpose, it might be possible to have a history with a natural plan for creatures who have a plan of their own" (ibid., 12).

12. See Edmund Husserl, *The Crisis of the European Sciences and Transcendental Phenomenology: An Introduction to Phenomenological Philosophy,* trans. David Carr (Evanston: Northwestern University Press, 1970), pp. 103–89 [Par. 28–Par. 54]. Husserl addresses what he terms Kant's limitation in knowing the substantial (four) causes of human knowledge of their own existence as he begins his discussion of "life world." This shows the contemporary reader how knowledge of the secular four causes is deepened every generation. What Husserl knew has been enlarged since his writing by minds such as Stephen C. Pepper and Hayden White, and my own inquiries and evidence based upon the avenues they have opened. Husserl writes of Kant: "Kant is certain that his philosophy will bring the dominant rationalism to its downfall by exhibiting the inadequacy of its foundations. He rightly reproaches rationalism for neglecting questions which should have been its fundamental questions; that is, it had never penetrated to the subjective structure of our world consciousness prior to and within scientific knowledge and thus never asked how the world which appears straightforward to us men, and to us as scientists, comes to be knowable a priori—how, that is, the exact science of nature is possible, the science for which, after all, pure mathematics, together with further pure a priori, is the instrument of all knowledge which is objective, i.e. unconditionally valid for everyone who is rational (who thinks logically).

But Kant, for his part, has no idea that in his philosophizing he stands on unquestioned presuppositions and that the undoubtedly great discoveries in his theories are there only in concealment; that is, they are not there as finished results, just as the theories themselves are not finished theories, i.e. do not have a definitive scientific form" (ibid., 103).

Only with such a horizon could efficient cause, material cause and formal cause (the series and sequence of actions needed to solve a formulated problem) be adequately discerned.

My methodology for discerning the fundamental principles of modernism and its progress into the present begins with Aristotle's four causes. These causes do not operate one after the other, or haphazardly. A thinker who seeks to address the nature of his or her life world (a concept I now employ) must integrate these four causes into the intentionality of his or her praxis as one who seeks to address the issues and problems of their functional existence. Aristotle knew this when he formulated a synthesis concept for the operation of the four causes in human thought and action—the concepts of Energeia and Ergon.[13] To strive to formulate and effect an end (final and formal cause) is Energeia. The end sought is an existent form in reality or its augmentation or replacement in imagination, which he termed the Ergon. We are as live, functioning creatures throughout our waking lives enacting the Energeia of our aims, and by that addressing the existent Ergon(s) of our attention which we wish to affect. In order to make these effects toward an augmented or new Ergon, we must comprehend our final cause, the formal cause—which indicates what activities we must pursue toward that end, know the materials involved, and by ourselves or with cooperation—the efficient cause—pursue our aims. In this metahistorical study of how individuals and groups sought to either sustain or change the Ergon of their life world, I will review how they understood it within the purview of the four causes. The phasal changes of such self-understanding and praxis by cohorts of a generation or more will be considered as the "metaparadigm" that linked these four causes to the several disciplines of a society. The changing knowledge through inquiry and praxis will generate an evolution of a metaparadigm to a

---

13. For the *ergon* is the telos, but *energeia* is the *ergon*, on which account the name *energeia* is drawn from ergon, and exerts all of its powers toward *entelecheia*.

Aristotle, Metaphysics, 1050a, 22

Ergon

"Let us again return to the good we are seeking, and ask what it can be. It seems different in different actions and arts; it is different in medicine, in strategy, and in the other arts likewise. What then is the good of each? Surely that for whose sake everything else is done. In medicine this is health, in strategy victory, in architecture a house ..."

Aristotle, *Nichomachean Ethics*, Book I, Ch. 7, 1097a, 15–20.

Aristotle says that if a thing has an *ergon*, "the *ergon* of each thing is its end" *(Eudemian Ethics* II 1, 1219a, 8) and he clarifies this by saying "the end is the best in the sense of being the last thing, for the sake of which everything else is done" *(Eudemian Ethics* II 1, 1219a, 10–11).*Energeia*"For each thing is capable of being sometimes actual, sometimes not, e.g. the buildable qua buildable; and the actuality of the buildable qua buildable is building. For the actuality is either this—the act of building—or the house. But when the house exists, it is no longer buildable; the buildable is what is being built. The actuality, then, must be the act of building, and this is a movement. And the same account applies to all other movements."

Aristotle, Metaphysics, Bk. XI, Ch. 9, 1066a, 1–7.

second metaparadigm. We will be examining four metaparadigms in the metahistory of Modernism from 1648 into the present.

Historical activity among our species is problem solving no matter what its guise. Each phase of a metaparadigm (there will be four), and each Modernist metaparadigm (there are four to this date) exhibits in human activity toward their respective life worlds attempts at problem formulation and problem solving. My address of history makes problem solving in its efforts and guises central to my narrative. The term "metaparadigm" was coined by Thomas Kuhn in the late 1960s.[14] His idea is also a core of my approach.

3

What is a metaparadigm?

The reader now must be informed of what I mean by the term "metaparadigm." Why are there four since roughly 1648? What is the nature of a "metaparadigm?" Why is it a useful concept in tracking the metahistory of knowledge? Historiography, that is the study of what conceptual organization is reflected in how an individual or group of individuals structure the events upon which they focus, calling it a "history" of the activity of a time, has always exhibited a "structure" that provides some sort of beginning, middle and end. Herodotus's appellation of *historia* to any study of such activity, be it political, social or other, meant that "inquiry" into time, place and manner was involved, but also its account was a "story" in the sense it gave human proportion to intent and outcome. Herodotus's initial sentence in his *Historia (Researches)* stresses the dual meaning of inquiry and story insofar as the human proportion of events:

> Herodotus of Halicarnassus, his *Researches* are here set down to preserve the memory of the past by putting on record the astonishing achievements both of our own and of other peoples; and more particularly to show how they came into conflict. (Herodotus, *Herodotus, the Histories* 1972, 41)

A "metaparadigm" generates such a concept in that it tracks how all inquiry of a time, in every field of knowledge, is affected by the normative concepts of that time, concept developed in separate disciplines or fields of knowing, yet sharing distinct characteristics. Moreover, it allows for the particular use of language and narrative form by researchers, the historiographical demand of Herodotus (and his successors in his time, most notably Thucydides). My above allusion to differing ways of treating cause and effect between Early Modernism and the emergence of Modernism in its divorce from religious cause and effect are such distinctions that govern how one inquires into and organizes the knowledge of human activity. But, a "metaparadigm" is both more and less than such an "epochal" change in knowing and explaining cause and effect. It is more insofar it tracks more finely how questions are asked and answered of a time. It is less insofar as it does not demarcate what I will call the epochal change between Early Modernism

---

14. Thomas Kuhn, *The Structure of Scientific Revolutions* (Chicago: University of Chicago Press, 1962).

and Modernism. An "epochal" change is used by me to denote a major cultural under-standing of what is sought as the final cause of knowledge. An "epochal" change is an identifier used by those who speak of how modernism emerged from the medieval, or how the classic age was ended by the thought of those who belonged to a Europe after the Fall of the Roman Empire.[15] We still are in an epoch that began roughly in 1648 and in which we still persist insofar the way the four causes of events are understood. In my conclusions I will speculate that our present metaparadigm foresees another 'epochal' change that will bring back a more complex final cause generated by the increasing glob-alization of civilization, an epochal change roughly equivalent to the move from sepa-rate Greek republics and democracies and Near Eastern dynasties to a world united by Hellenism, and then the Roman Empire.

What links these four "metaparadigms" together in this epoch of Modernism is the general characteristic of the epoch—problems are posed by humans with the under-standing that they themselves are responsible for how the formulation and address of the problems can lead to an answer. This is the secular final cause of how we as a species in the West normatively pursue the issues of daily and long-range existence within the dis-ciplines of the arts and the sciences. My thesis in this text is that all intentional thought within the disciplines of the arts and the sciences have this focus upon the individual inquirers of a discipline being consciously responsible for formulating problems, pur-suing the implications for research of such a formulated problem, and attending to the solutions indicated by putting them into effect. A "metaparadigm" I will demonstrate has four phases—each a certain period of years—in which:

(1) a major problem is posed in a discipline;
(2) the long range significance of such a problem to be solved is articulated;
(3) results begin to be demonstrated in their useful effects, albeit challenging existing ways of pursuing knowledge;
(4) there is an integration of what preceded the new problem formulations and what has been gained as new, effective knowledge in guiding the activities of disciplines and culture.

We are now in what I will describe as the fourth Modernist metaparadigm, and on the verge of leaving the third phase of it and entering the fourth. Yet, this fourth metaparadigm in its dating, roughly, 1970–2050, is called by many post-Modernism. Post-Modernism has in almost every discipline of the arts and the sciences brought into question the Modernist claims to an unbroken progress in our self-knowledge and the knowledge of our world. I will counter this negative orientation, which I will show to have positive, functional aspects that are in themselves progress. Charles Jencks's appella-tion that replaces the term "post-Modernism" with that of "critical Modernism" makes many of arguments I will make.[16]

---

15. https://www.oed.com/view/Entry/63645?redirectedFrom=epoch#eid,
16. Charles Jencks, *The Story of Post-Modernism, Five Decades of the Ironic, Iconic and Critical in Architecture* (Chichester, West Sussex, UK: Wiley, 2011), 91, 101, 117, 125, 139, 141, 148, 150, 157, 158, 225.

Wilhelm Dilthey (1833–1911) is the first historiographer who spoke of phases of all disciplines of the arts and sciences of a time that shared a world view, and developed concepts that guided their inquiry and explanation informed by this world view. William Kluback synthesizes Dilthey's theory of generational change in the service of the cultural systematization of new ideas and values in his study of Dilthey's historical thought:

> He sketched the pattern of development (of new ideas and values in culture) in four stages. In the beginning a new outlook grew primarily out of a new life-relationship which no longer fitted within the old categories. The new life relationship then expressed itself in new concepts and in fragmentary systems in poetry and unsystematized thought. Out of the early studies there grew up comprehensive, systematic metaphysical constructions. Finally, the new world-view reached maturity when critical investigations laid the epistemological bases for these systems. Only in this final stage, said Dilthey, could lasting progress be made toward understanding the phenomena of historical reality.[17]

While Dilthey did not speak of the stages of "problem-solving" per se, he did examine how the disciplines of a time in the arts and the sciences adapted old concepts and coined new concepts as the bases of their inquiry and explanation. Dilthey did not specify how long the distinct phases of such a change in world view lasted. Moreover, only his fourth phase provided a solid system of inquiry based upon epistemologically sound methods of inquiry and explanation insofar as what I am referring to as the Aristotelian four causes.

Dilthey's conception of the four phases in the development of a "world view" that generates new understandings and concepts in the arts and the sciences of a time is in itself a progression of Western knowledge. His thought occurs at the beginning of the third metaparadigm of knowledge, as I will show, in the late 1860s and 1870s. A century earlier, in the late 1760s, Johann Herder had coined the term "Zeitgeist" ("spirit of the time") to discuss how across the culture of that time, certain ideas and values were shared, including in what, how and why inquiry and explanation was pursued.[18] He did

---

17. Dilthey's theory of staged cultural systematization is encapsulated "Der Entwicklungs-geschichtliche Pantheismus nach seinem geschichtlichen Zusammenhang mit den llteren Pantheistischen Systemen," in Wilhelm Dilthey, *Gesammelte Schriften*, Vol. II (Stuttgart: B. G. Teubner, 1957), 312–14. The entire second volume *Weltanschauung und Analyse des Menschen Seit Renaissance und Reformation*, however, is the cultural history of the genesis of the modern age. William Kluback, *Wilhelm Dilthey's Philosophy of History* (New York: Columbia University, 1956), 38. Kluback refers to Dilthey's essay "Der Entwicklungsgeschichtliche Pantheismus nach seinem geschichtlichen Zusammenhang mit den allteren Pantheistischen Systemen," in Wilhelm Dilthey, *Gesammelte Schriften*, Vol. II (Stuttgart: B.G. Teubner, 1957), 312–14. Further references to Dilthey's own writings are to this edition of his *Gesammelte Schriften* in multiple volumes (each volume with its own publication date).

18. See Johann Herder, "Über die Reichsgeschichte, Ein historischer Spaziergang," in *Zur Philosphie der Geschichte, Eine Auswahl in Zwei Bänden* (Berlin: Aufbau Verlag, 1952), 263. See also, Johann Herder, "Auszug aus einem Briefwechsel über Ossian und die Lieder alter Völke," in *Von deutscherArt und Kunst*, ed. Edna Purdie (Oxford: Oxford at the Clarendon Press, 1924), 120–21.

not speak of phases of a "Zeitgeist," and thus the deepening of the concept by Dilthey. Indeed, my present work with this text can be viewed as a further progress in this address of a cultural period within which the disciplines of the arts and the sciences shared methods of inquiry and explanation. I differ with Dilthey's insights in this respect. I find that each of the four stages generates coherent views of historical reality, not solely in the fully developed cultural-historical ideas of Dilthey's fourth phase that guide long-term historical life.

Dilthey did link to his concept of the "world view" a more focused specifier that is a ground for my more fully developed concepts for each of the four phases I demarcate. That was his notion of "world representations or theories" (Weltbilder) that stemmed from the universal world view of the time. I will show how certain formulations such as axioms or more developed theories were generated respectively in each of the four phases I demarcate that stem in their phasal differences from a universally accepted world view. My use of Aristotle's four causes, the final, formal, material and efficient, as well as other guiding concepts that stemmed from them, will be my application of Dilthey notion of "world representations and theories" (Weltbilder) in each of the four phases of a metaparadigm.

There are several major contributors to the historiography of this fourth metaparadigm of Modernism whose thought informs my own approach. Most notably, of course, is Thomas Kuhn who coined the term "metaparadigm." A paradigm for Kuhn is the equivalent of the Diltheyan Weltbild that governed the conduct of inquiry in a discipline of a generation. Kuhn's "metaparadigm" would be the assumptions that governed problem or hypothesis articulations, methods of inquiry into the problem or hypothesis formulations, and the favored principles that provided explicative and explanatory expositions of the conduct of inquiry in all the disciplines of the arts and sciences of a time. Kuhn does break down the concepts of paradigm and metaparadigm into manageable concepts that can be illustrated with methods of inquiry and explanation. The paradigm as Weltbild is not philosophical, rather tangibly present in workable hypotheses, methods for gathering empirical evidence, and responsible in its rational explications and explanations to the rules of logic, even new logical formulations capable of demonstrating new insights. Kuhn explicates this quite clearly. He dwells on the tangibles of problem formulation, method of inquiry, explications and explanations in his *The Structure of Scientific Revolutions*. One of Kuhn's interpreters, Margaret Masterman, makes a strong case for this pragmatic, procedural focus of the Kuhnian paradigm.[19] She speaks of the "paradigm" in Kuhn's "concrete" application: "as an actual textbook supplying tools, as actual instrumentation" of how science was practiced according to normed procedures in one or more generations (Masterman 1970, 65).

Kuhn articulates the concept of a metaparadigm several times in his *The Structure of Scientific Revolutions* according to Margaret Masterman, who uses this term for the

---

19. Margaret Masterman, 'The Nature of a Paradigm," Criticism and the Growth of Knowledge, *Proceedings of the International Colloquium in the Philosophy of Science,* London, 1964, Vol. 4, ed. Imre Lakatos and Alan Musgrave (Cambridge: Cambridge University Press, 1970), 59–90.

first time in describing Kuhn's theory in the above-cited 1964 colloquium dedicated
to Kuhn's work. Masterman says that Kuhn saw not only change over time in the
basic assumptions of a discipline, but in a broader world-view shared more generally across a scientific culture: "For [Kuhn's] metaparadigm is something far wiser
than, and ideologically prior to, theory: i.e. a whole Weltanschauung" (*The Nature
of a Paradigm,* Masterman 1970, 67). Masterman states that one can read of the
metaparadigm, rather than the mere paradigm on many pages in *The Structure of
Scientific Revolution* —pp. 2, 4, 17, 102, 108, 117–121 and 128 of the first edition *(The
Nature of a Paradigm,* 1970, 65).[20] Kuhn speaks, for example, according to Masterman,
of the metaparadigm shared across many fields because of the "shift of vision" generated by a crisis in cultural understanding that effects over time a new cultural systematization of science. The crises recur over centuries introducing new ideas and values.
Such a crisis occurred according to Kuhn in the late sixteenth and early seventeenth
centuries. Harbingers of the new cultural vision were, for Kuhn, Galileo, Descartes
and Newton:

> Aristotle and Galileo both saw pendulums, but they differed in their interpretations of what
> they had seen.
>
> Let me say at once that this very usual view of what occurs when scientists change their minds
> about fundamental matters can be neither all wrong nor a mere mistake. Rather it is an
> essential part of a philosophical paradigm (read according to Masterman "metaparadigm")
> initiated by Descartes and developed at the same time as Newtonian dynamics. That para-
> digm has served both science and philosophy well. Its exploitation, like that of dynamics itself,
> has been fruitful of a fundamental understanding that perhaps could not have been achieved
> in another way. But as the example of Newtonian dynamics also indicates, even the most
> striking past success provides no guarantee that crisis can be indefinitely postponed. Today
> research in parts of philosophy, psychology, linguistics, and even art history, all converge to
> suggest that the traditional paradigm is somehow askew (my emphases). (*Structure of Scientific
> Revolutions,* Kuhn 1962, 119–20)

Kuhn's 1962 insight into what he sensed was changing in a profound manner I see as the
notion begun with Descartes that there is one objective truth in every facet of nature—
what I will call univocal objectivity. He sensed the beginning of a "multiple objectivity,"
where various arguments were valid addresses of the same phenomena. His research
gives a new ground to objectivity in that sense. What becomes "post-modernism" is at
work in its questioning of "univocal objectivity" by the late 1960s.

Kuhn is best in reviewing the "concrete" level of the "paradigm" that guides a discipline within a metaparadigmatic set of assumptions. One can abstract at least 21 "concrete" characteristics of the procedures of disciplines that erect an emergent new
paradigm and then establish it as a normed conduct of inquiry in Kuhn's *The Structure of
Scientific Revolutions:*

---

20. Thomas Kuhn, *The Structure of Scientific Revolutions* (Chicago: University of Chicago Press, 1962).

Characteristics of an emergent paradigm

(1) New paradigms are conservative: they preserve most of the terminology, examples, and methods of previous views (1962, 7, 141–42, 168).

(2) The emergence of a new paradigm is a slow process (where new problematic insights and methods incrementally replace traditional procedures among an increasing body of inquirers) (1962, 84–85).

(3) The emergence of a new paradigm is accompanied by the creation or adaptation of new methods for collecting and analyzing data (1962, 84–85).

(4) New paradigms explain existing anomalies (better, more simply, more completely than did previous views) (1962, 154).

(5) In new paradigms explanation generally runs ahead of proof-the details which provide firm support for arguments must be filled in (1962, 44, 46).

(6) A new paradigm appeals to others (i.e., those working in other fields) because it offers illumination and confirmation of their assumptions and insights that have begun to changes their paradigms) (1962, 120, 166).

(7) Popular acceptance of a new paradigm requires that it be consonant with deeply held self-evident beliefs (i.e., with the metaparadigm which exists at the time) (1962, 127).

(8) **A** new paradigm must be seen as culturally useful if it is to stimulate "normal" scientific activity (1962, 23–24).

Characteristics of a normalized paradigm

(9) A paradigm is characterized by the use of distinctive methods-methods accepted by all those carrying out research within the paradigm (1962, 47–48).

(10) A paradigm produces a standard stock of demonstrations and examples used to illustrate basic methods and prove basic principles (1962, 46–47).

(11) A paradigm recognizes a limited number of valid sources for data (1962, 4, 24).

(12) A paradigm insulates researchers from distracting problems and phenomena (1962, 37).

(13) A paradigm guides researchers in choosing activities and problems for investigation (1962, 24).

(14) A paradigm limits the range of hypotheses which can be offered as explanations for problems (1962, 24).

(15) Those sharing a paradigm uses uniform language (1962, 127–28, 135).

(16) Those working within a paradigm must meet rigid standards for orthodoxy (i.e., to be accepted as members of the community of researchers) (1962, 167–68).

(17) Researchers within a paradigm never question its foundations-its basic hypotheses, methods, categories of acceptable evidence, and so on (1962, 47).

(18) Researchers sharing a paradigm are often unable to articulate the foundations on which the paradigm is built (1962, 47).

(19) A paradigm produces data which lasts even when the paradigm is replaced as the previous concepts and methods when rigorously applied generated actual facts (that the new paradigmatic methods might ignore) (1962, 139–42).

(20) Most paradigms use books only as texts for teaching new researchers; active communication between or among established researchers is carried out through highly specialized short reports (1962, 135–37).

(21) All of the paradigms which are widely accepted in any period by a culture share many common features of problematizing, methodology and acceptable evidence (1962, 120).

Thus, while Kuhn does not refer to phases of the emergent and finished metaparadigm that consists of its member paradigms, his characteristics enable me to speak of four phases that can be identified in each disciplinary paradigm as well as the metaparadigm as a whole.

In discerning the four phases of a metaparadigm, two other thinkers who can be counted as contributing to the fourth Modern metaparadigm must be accounted for. Stephen C. Pepper in his *World Hypotheses, Prolegomena to systematic philosophy and a complete survey of metaphysics* has discerned four patterns of the structuring of events that I have found to be central to aspects of the four phases of metaparadigm development. Thinkers such as Pepper are ahead of their time by often more than a decade—as I will argue in my text.

Hayden White, who uses Pepper's *World Hypotheses* in his application of the idea of "metahistorical" phases of Western cultural history, is closest to my own approach to the concept of an evolving metaparadigm. White shows in his 1973 study *Metahistory: The Historical Imagination in Nineteenth-Century Europe* phases of emergent metaparadigms that evidence the use of distinctly different types of narrative grammar. His study will greatly inform how I discuss the narrative structures used in the four phases which I present.

4

What are the four phases of a metaparadigm?

The cultural systematization of new ideas and values in the West is a fourphased cycle. The initial phase is an emergent new idea and/or value; the second phase is a model for all the significant individual and institutional changes; the third phase is a pragmatic deconstruction of the older system and a construction of the new one; the fourth phase is a normalization of the new system so that it becomes a habitual right order for all its participants. There are four categorical assumptions which underlie the development and implementation of cultural activity in the first three phases that precede the fourth phase of normalization:

(1)   the complex individual differences of all integrities (person, place or thing) that participate in an assumed common totality;

(2)   the freedom of intelligent beings or accident/chance for things without mind or will;

(3) quantum change in states-of-affairs;
(4) mind rather than matter as the chief content of judgment.

The fourth phase presents the converse of these concepts: one sees

(1) the collective expression of persons, places, and things, that is, having the totality
     formed by them in common by dint of shared characteristics;
(2) determinism in the laws of function in intelligent beings and in things without mind
     or will;
(3) a conception of duration of states-of-affairs;
(4) matter rather than mind is the chief content of judgment.

The efforts of the fourth phase are integrative of all current and traditional approaches,
appealing to the common-sense adherence to what seems to have always been true in
methodology, evidence and explanation with the augmentations of the new approaches.
One can appreciate this implicit, determinative set of understandings with Ludwig
Wittgenstein's deconstructive axiom:

> A picture held us captive. And we could not get outside it, for it lay in our language and
> language seemed to repeat it to us inexorably.[21]

Each phase of a metaparadigm will have a different Aristotelian cause as an emphasis.
The semantics of the thought will reflect the significance of that cause that dominates
each phase.

The emergent reality of a new cultural systematization has a final casual emphasis,
indicating new problematics are needed to address the particular societal needs seen to
be the case. The very grammar is replete with metaphors, synecdoches or more formal
axioms are establishing new ways to comprehend the issues addressed. Nature in its established laws also becomes a new totality in its depiction. This first phase thus generates
claims that will be examined in the next phase in a more systematic manner.

The concepts developed in the second phase are linked together systematically. The
formal cause is the dominant thought in its care in systematically creating a menu for
inquiry to validate a conceptual thoroughness. Their veracity is being carefully established in the inquiry of the third phase. The Aristotelian material and efficient causes
are the hallmark of the third phase. The third phase is that of everyday change in its
pragmatic work. One does not look at the mountain—to put it in a Taoist analogue—
rather, to each step of the conceptual path. There is much conflict as the results of
the third phase challenge existing understandings in how problems in that area of concern are to be solved. Out of this conflict-filled, pragmatic reality the fourth phase is
opened to compromise with what all members of the society share insofar as a world

---

21. Ludwig Wittgenstein, *Philosophical Investigations*, 3rd ed., trans. G.E.M. Anscombe (New York:
     Macmillan, 1958), 48e (Prop. 115).

vision. The Aristotelian final cause and formal cause create horizons for inquiry, explanation and the rigor of the material and efficient causes of praxis in this fourth phase—the
longest phase of the four phases of the metaparadigm.

The political conflicts of the Tory politician Lord Bolingbroke in the third phase will
be muted in his royalist theories of the fourth phase of the First Modern metaparadigm,
giving us the first formulation of the "balance of power" between those who favored
dynastic executive authority and those who wanted the voice of elected representatives.
From his losing effort to have a Royalist society—the conflict driving him into exile, he
enabled the compromise of the fourth phase that sought cooperation among the whole
populace with the aristocratic tradition of English rulers. In his *The Patriot King* (1737) we
hear this plea for common-sense compromise:

> Since men were directed by nature to form societies, because they cannot by their nature
> subsist without them, nor in a state of individuality. And since they were directed in like
> manner to establish governments, because societies cannot be maintained without them, nor
> subsist in a state of anarchy; the ultimate end of all governments is the good of the people, for
> whose sake they were made, and without whose consent they could not have been made. In
> forming societies, and submitting to government, men gave up part of their liberty to which
> they are all born, and all alike. But why? Is government incompatible with a full enjoyment
> of liberty? By no means. But because popular liberty without government ill degenerate into
> licence, as government without sufficient liberty will degenerate into tyranny, they are nec-
> essary to each other, good government to support legal liberty, and legal liberty to preserve
> good government.[22]

Lord Bolingbroke in the period of time spoke with the Baron de Montesquieu in
France,[23] and from those conversations we get the concept of the "balance of power"
in Montesquieu's 1748 *The Spirit of the Laws*,[24] the essential compromise that becomes
the basis of the US government (in a second phasal development in American thought).

The fourth phase is that of dwelling perforce in a collective reality common to all.
Time is seen to change gradually, being measured in terms of an institutional development common to all. Material realities surround one, evidence of the tangibility of the
"right order"—a seemingly immovable order—that has been established. These "meta"
assumptions find their way into disciplinary expression and popular culture in the years
involved with each phase.

From a systems theory perspective, the Diltheyan intergenerational sequence that
I have modified and presented can be considered an "organismic analogy." Human
beings require a stable context of beliefs, assumptions and the organizational forms
that maintain these perspectives. There is a balance of mind and body in any culture.

---

22. Henry St. John, Lord Bolingbroke, "The Patriot King," in *Bolingbroke's Letters on the Use and Study
of History, etc.*(London: Ward, Lock & Co., N.D), 204.
23. https://ouclf.iuscomp.org/montesquieu-in-england-his-notes-on-england-with-commentary-
and-translation-commentary/
24. See Montesquieu, *The Spirit of the Laws*, trans. and ed. Anne Cohler, Basia Miller and Harold
Stone (Cambridge: Cambridge University Press, 1989), 31–36 [Book I, Chapters 1–5].

As Dilthey understood, ideation and biology are separate realms, but the nature and pace of ideation accommodate physical needs, and is best imputed in a natural rhythm of human information-processing. The four-phased introduction, modeling, development and stabilization of a new cultural system through the arts and the sciences and the popular culture are a "self-evident" process, if one considers how the *Energeia* of human praxis strives to alter the current *Ergon* in each area the practitioners of the society address. Ludwig von Bertalanffy quotes M. Haire in this regard:

> The biological model for social organizations—and here, particularly for industrial organizations—means taking as a model the living organism and the processes and principles that regulate its growth and development. It means looking for lawful processes in organizational growth.[25]

The gradual integration of new ideas into models, and then actual change of the culture through an application of those ideas requires several decades to accustom a populace to the significance of what is afoot. The work of deconstruction and reconstruction that reorganizes institutions can be traumatic, and requires consensus and accommodation. The new ideation that seems sufficiently significant to reach public effect undoubtedly resolves issues that ease the populaces that can see the value. Then, the establishment of a durative "right order" that serves as the normal life and standard of right and wrong develops that as a final phase continues for another generation (approximately 30–40 years) before the culture again engages in large-scale reexamination of its life issues. Ortega y Gasset's view of what a generation of inquirers is, and the emergence of a stress on individual realities vis-à-vis collective realities, has aided my own formulation of the sequence and length of the phases of a metaparadigm. See Ortega y Gasset on these issues (*Gesammelte Werke* 1950, 1: 533–37; 2: 112–15, 2:455–60; and, 2: 462–66).

The temporal duration of each of the four phases is an interesting issue. I have found that a metaparadigm has an existence as an emergent and normalized world view for approximately 90–110 years, depending upon external events, primarily the cultural disruption of war. The first phase is one of novel ideation, unproven, but sufficiently provocative to impel the development of the systematic conceptual architectonic of the second phase. The first phase then is generally is one of personal insights, often shared by others. Research has shown me that it endures approximately 10–15 years, even though some "geniuses," and one must give them this honor, foresee this phase in their fourth phasal work, and, must be seen as an exception to the rule insofar as its far earlier articulation. But, here we deal with scattered individuals, and not a well-populated first phasal norm for individuals addressing disciplinary problems. An example of that is the thought of Stephen C. Pepper in the 1930s and 1940s, in the fourth phase of the third Modern metaparadigm. His articulation of four hypotheses for structuring the world were each never demonstrated evidentially, although the axiomatic conceptual rules that developed each "world hypothesis" were

---

25. See Ludwig von Bertalanffy, *General System Theory,* revised edition (New York: George Braziller, 1968), 112.

valid structural insights that can guide a second phasal systematic architectonic, and guide exhaustive material evidence to validate their claims. Pepper's "world hypotheses" were also in service of a more finely integrated world-in-common, a view that while showing differing ways of structuring events, insisted on one in-common objectivity realized through complementary trains of thought. In that he was a member of the fourth phasal realities of his time. When his insights were taken up by Hayden White in the late first phase and second phase of our present metaparadigm, their trajectory was not complementary integration, rather differing periods of cultural history dominated by one of the world hypotheses or another. I agree with White, and will demonstrate that in my analyses which draw heavily from Pepper's four "world hypotheses." Each of the "world hypotheses" undergirds a different phase of a metaparadigm's four phases. Each enables us to see what is the dominant event structure for the culture in that time, and why.

The second phase of the emerging metaparadigm ordinarily is about 15 to 20 years. In this phase, the creative mind who initiated the new ideation in the first phase works to develop all the implications in his or her field. There is more evidence, of course, for the new developing architectonic of ideas more and more systematically complete as a thorough approach to one's discipline, but not such that the traditional approach changes for most inquirers. Bolstering the efforts of the original thinker, and those who begin to follow him, is the insight of Immanuel Kant on how we discover the implications of our original flashes of insight:

> It is unfortunate that only after we have spent much time in the collection of materials in somewhat random fashion at the suggestion of an idea lying hidden in our minds, and after we have, indeed, over a long period assembled the materials in a merely technical manner, does it first become possible for us to discern the idea in a clearer light, and to devise a whole architectonically in accordance with the ends of reason. Systems seem to be formed in the manner of lowly organisms, through a *generatio aequivoca* from the mere confluence of assembled concepts, at first imperfect, and only gradually attaining to completeness, although they one and all have had their schema, as the original germ, in the sheer self-development of reason. (Kant, *Critique of Pure Reason*, 1965, 655 [A 835, B 863])

I will venture to state that what Kant saw in his own self-development of ideation, and those of his time and before in the years of Enlightenment was the full complement of Aristotle's four causes being fulfilled in "the end of reason." It is well-known that Kant constructed his own a priori causal ideas being informed by Aristotle's causes (Kant, *Critique of Pure Reason*, 1965, 114 [A 81, B 107]; 281 [A 269, B 325]). The time needed for this second phasal architectonic then requires such sufficiency of ideation. Kant's own architectonic of the second phase he belonged to of the late Enlightenment was from approximately 1770 through 1787. His Inaugural Dissertation of 1770, *De Mundi Sensibilis atque Intellibilis Forma et Principiis (Concerning the Form and Principles of the Sensible and Intelligible World)* appears to be the first efforts to establish what became a finished architectonic of "pure reason" by 1787.[26] Kant will be included in our discussion in Part II, Chapters Six

---

26. See *The Stanford Encyclopedia of Philosophy*, https://plato.stanford.edu/entries/kant/

and Seven—the second and third phases of the second Modern metaparadigm, charting his own "genius" ideation, and the successive phases of his thought in tandem with those who shared inquiry and explanation in other disciplines of this time.

This is a "spiral" return at a higher level to each phase as one metaparadigm evolves into another. This reorientation of every phase from metaparadigm to metaparadigm indicates the value of every phase of sound research over the generations of Western knowing. The idea of "spiral return" is true for each of the four phases, whose new insights in the developing "world view" have a foundation in the previous phase of the older metaparadigm. This gives new ideation a foothold, using the older language of the previous metaparadigm's specific phase, be it the first, second, third or fourth, as a construct to reconceptualize it in the present phase. Giambattista Vico spoke of the "spiral" return of problematics from age to age in his 1725 *The New Science*. Moreover, his saw this spiral return as a deepening of understanding into the kind of problematics inquirers of the new age took up from that past.[27] Thus, we will see in this understanding by Vico a "metahistorical" insight in the first metaparadigm of Modernism's fourth phase that addressed collective realities in a manner that set a foundation for the new social science in Comte and Marx, and the spiral return of other phases in such augmented and more complex understandings.

5

What historical disciplines do I offer to demonstrate my metahistorical approach, and what individuals do I choose to be the discipline's spokesperson?

In choosing individuals and disciplines to demonstrate the above phases with their characteristic thought, I have sought individuals whose ideation participated in as many of the phases as possible. That was not always possible because of the duration of mature thought, which ordinarily begins in one's twenties and possibly into his or her seventies and eighties as we move toward the present. Kant, for example, born in 1724 will die in 1804 before the fourth phase of his metaparadigm, and thus never demonstrate a move toward cooperative, pragmatic, self-limitation in his customary challenge to what exists around him. I will begin the text with Thomas Hobbes, who was born in 1588, and began to write in the fourth phase of the Early Modern. He will live until 1679, but his contribution to philosophy and history will end in the 1660s, a life of thought that will end in the second phase of the first Modern metaparadigm. Johann Wolfgang von Goethe, who will represent Literature in my text for the second Modern metaparadigm, was born in 1749 and died in 1831, and will demonstrate the characteristics of all four phases of that metaparadigm in his career of thought and literature.

---

27. See Hayden White's emphasis of this higher understanding with the spiral return in his essay "Croce's Criticism of Vico," in Giambattista Vico, An International Symposium, ed. Giorgio Tagliacozzo and Hayden White (Baltimore: The Johns Hopkins Press, 1969), 385.

I will track several disciplines throughout the Modern metaparadigms, history/philosophy of history/historiography, literature, the logic of calculus, biology, and in the present metaparadigm, psychology, micro-sociology and group dynamics. The harder lines of disciplinary exclusiveness begin in the late Enlightenment, and, the social sciences only in the mid- and late nineteenth century. I will focus upon one or several thinkers in each of these disciplines in each phase of a metaparadigm. Other example thinkers will appear in each of the four phasal essays, but not in the depth of the chief mind(s) in that discipline in that phase.

Part I

The First Modern Metaparadigm, c.1648–c.1750

Ever since man has been able to reason, philosophers have obscured the question of free will; but the theologians rendered it unintelligible by absurd subtleties about grace. Locke was perhaps the first man to find a thread in the labyrinth, for he was the first, who, instead of arrogantly setting out from a general principle, examined human nature by analysis. For three thousand years people have disputed whether or not the will is free. In the *Essay on Human Understanding*, Locke shows that the question is fundamentally absurd, and that liberty can no more belong to the will than can color and movement.

What is the meaning of this phrase "to be free?" It means "to be able," or else it has no meaning. To say that the will "can" is as ridiculous at bottom as to say that the will is yellow or blue, round or square. Will is wish, and liberty is power. Let us examine step by step, the chain of our inner processes without befuddling our minds with scholastic terms or antecedent principles.

Voltaire, Philosophical Dictionary c.1750

Voltaire formulates the most significant characteristics of the first Modern metaparadigm in his reflections on Locke and "free will." The four phases of this metaparadigm will introduce heightened methods of self-analysis, methods for the analysis of human nature and the natural world, and separate this gain of human understanding and doing from the intervention of the Divine. This is a new epoch in human self-development. From its onset, we will see the view of a sovereign human being, whose very being is to be autonomous in his or her choices, indeed finding, as Hobbes will, that this "free will" is the major purpose of the age, an age that will reconstruct society's vision of authority, law, learning and moral action. As Martin Heidegger was to say in a later Modern metaparadigm, referring to great epochs such as that of ancient Greek thought and its ways of conceiving the human person and his community "But what is great can only begin great. Its beginning is in fact the greatest beginning of all."[1] He also saw such greatness as a potential within the beginnings of this epoch that begins in the mid-seventeenth century.[2]

---

1. Martin Heidegger, *An Introduction to Metaphysics*, trans. Ralph Manheim (New York: Anchor Books, 1961), p. 13.
2. Martin Heidegger, "The Question Concerning Technology," in *The Question Concerning Technology and Other Essays*, trans. William Lovitt (New York: Harper Torchbooks, 1977), pp. 22–23.

Chapter One

THE FIRST PHASE: SEMINAL IDEATION, C.1648–C.1670: THE FOCUS UPON DEFINITION AND HYPOTHESIS

All these motives, I am suggesting, could take enough of a hold over some souls to produce some actions and effects which might seem supernatural without actually being so.

Sir Kenelm Digby's (1603–1665) letter to the Louis de Rohan (1598–1667), prince de Guémené, concerning the possessed people at Loudon, c. Fall, 1636 (attached to a letter sent to Thomas Hobbes, January 17, 1637).[1]

In your Logike, before you can manage men's conceptions, you must show how to apprehend them rightly; and herein I would gladly know wither you work upon the general notions and apprehensions that all men (the vulgar as well as the learned) frame of all things that occurred unto them; or whither you make your ground to be definitions collected out of a deep insight into the things themselves.

Sir Kenelm Digby's letter to Thomas Hobbes, January 17, 1637, P. 42 (Letter 25).

If the Society would retreat a little from its Baconian stance, and accept that it might be permissible to gather information with a specific aim in mind, rather than collecting material on every conceivable subject, the work might be completed more quickly. "I mention this, to hint only by the by, that there may be use of Method in the collecting of Materials, as well as in the use of them, and to show that ... there ought to be some End and Aim, some pre-designed Module and Theory, some Purpose in our Experiment".

Robert Hooke (1635–1703), written on a lecture on Earthquakes in 1667 or 1668.[2]

These epigraphs introduce the transitional shift from the fourth phasal Early Modern principles of knowing (c.1620–1648) outlined in the Introduction to what becomes the first phasal, First Modern metaparadigm principles of knowing (c.1648–1672):

Fourth phase (c. 1620–1648) of the Early Modern Metaparadigm Principles of Knowing

(1)   the collective expression of persons, places, and things, that is, having the totality formed by them in common by dint of shared characteristics

---

1. *The Correspondence of Thomas Hobbes*, Volume I: 1622–1659, ed. Noel Malcolm (Oxford: Clarendon Press, 1997), 47 (Letter 25).
2. Stephen Inwood, *The Man Who Knew Too Much: The Strange and Inventive life of Robert Hooke 1635–1703* (London: Macmillan, 2002) 124.

(2)   determinism in the laws of function in intelligent beings and in things without mind or will;

(3)   a conception of duration of states-of-affairs;

(4)   matter rather than mind is the chief content of judgment.

First through Third Phases (c. 1648–1720) of the First Modern Metaparadigm Principles of Knowing

(1)   the complex individual differences of all integrities (person, place, or thing) that participate in an assumed common totality

(2)   the freedom of intelligent beings or accident/chance for things without mind or will;

(3)   quantum change in states-of-affairs;

(4)   mind rather than matter as the chief content of judgment.

Kenelm Digby was in the same quandary many others of his time. They had a solid belief in Divine final cause, but under the influence of other believers, such as Descartes, Grotius and Pascal, as well as Hobbes, began to question to what extent individuals were free to effect causal change themselves. The best Early Modern thinkers became inclined toward the idea of natural law being not only known by ordinary persons through inquiry, but used by their human designs to solve the problems of inorganic, organic and human nature that confronted their "life world." This began to challenge the problem formulation and problem-solving principles of the fourth phase. The fourth phase of any metaparadigm is prone to a determinism in that an in-common set of beliefs seems to be a gravity that all struggle to share without conflict, rather than challenging it constantly as had been done in the earlier decades of the metaparadigm. New insights were less stressed than working for the in-common world with what was known. This is the Baconian model of science. Continue to observe the natural world, but in an encyclopedic way with the methods that exist. We will see the method of scientific inquiry in the Baconian model of inductive inquiry that eschews theories with their supporting hypotheses, relying on traditional understandings of cause and effect (formal, material and efficient) that Hooke objects to will be a constant conflict between fourth and first phases of metaparadigms, instilled in the time-frame of human problem-solving.

An Early Modern theologian and philosopher, Francisco Suarez (1548–1617) wrote voluminously in both the "secondary cause" that emanated from God which humans could affect, and, the "proximate cause" that was wholly in an accessible human "life world."[3] This definitional broadness enabled the shift we will see to a wholly accessible natural world without considering God's role in the first Modern metaparadigm. Becoming aware of these distinctions enabled a freedom for inquiry, yet also the demand for formulating hypotheses based upon careful definitions of conceptions and terms. Rather than a trust of and reliance upon the fourth phasal understanding of collective reality, while

---

3.   Francisco Suarez, *Suarez on Individuation, Metaphysical Disputation V, Individual Unity and Its Principle*, trans. from the Latin with Introduction, Notes, Glossary, and Bibliography by Jorge J.E. Gracia (Milwaukee, Wisc.: Marquette University Press), 188–89.

themselves giving testimony to that reality in some ways, these thinkers became more oriented to the singular individualism that was their responsibility. Individuals such as Digby and Hooke were challenged to "become aware of being in new ways." This knowledge will spur a transformation of mind that in thinkers, such as Hobbes, to make the break with final cause in life being ascribed to the Divine. Leibniz, on the other hand, will allow God as the first (or final) cause, but his emphasis in his thought will be on "secondary" and "proximate" causes (formal, material and efficient).

A. History/Philosophy of History

Thomas Hobbes (1588–1670) and Samuel Pufendorf (1632–1694)

In his 1641 work *De Cive (On the Citizen)* Hobbes begins with the collective horizon of argument that is typical of the fourth phasal duration of a metaparadigm, in this case that of the Early Modern. He writes of man and society (in a manner that can be likened to the formulations of Bolingbroke and Montesquieu a century later, in their fourth phasal thought):

> Since we see that men have in fact formed societies, that no one lives outside society, and that all men seek to meet and talk with each other, it may seem a piece of weird foolishness to set a stumbling block in front of the reader on the very threshold of civil doctrine, by insisting that man is not born fit for society. Something must be said in explanation. It is indeed true that perpetual solitude is hard for a man to bear by nature or as a man, i.e. as soon as he is born. For infants need the help of others to live, and adults to live well. I am not therefore denying that we seek each other's company at the promptings of nature. But civil Societies are not mere gatherings; they are Alliances [Foedera], which essentially require good faith and agreement for their making.[4]

Hobbes articulates several things in the above that are of fourth phasal character:

(1)   He reaffirms what emerged in Early Modernism implicit in the thought of Jean Bodin,[5] and more explicit in the thought of Hugo Grotius,[6] that the person is a member of society according to the law of nature.

(2)   The person, however, must become aware of how is everyday adult being has a learned duty to the model of the society and governance of his or her time. This is "being aware of being aware" and becomes the basis of Hobbes' conception of a "social contract." In this thinking he inherits the Early Modern advance in reflective thought.

---

4.   Thomas Hobbes, *Hobbes, On the Citizen*, ed. and trans. Richard Tuck and Michael Silverthorn (Cambridge: Cambridge University Press, 1997), 24.

5.   Jean Bodin, *On Sovereignty*, ed. and trans. Julian H. Franklin (Cambridge: Cambridge University Press, 1992).

6.   See Hugo Grotius, *The Rights of War and Peace, Including the Law of Nature and of Nations* (New York: Cosimo, 2007), 17–30 [Book I, Chapter I].

(3)   Although formulated as a fourth phasal insight, Hobbes speaks of the imperfect nature of the human for civil society, stressing rather a self-conscious overcoming of that trait. This focus upon the necessity of a reasoned "free will" will become a major principle through the Modernism that follows. Hobbes will focus only a decade later (1651) as a major voice in the new ideation of the first Modern metaparadigm, the individual's nature and role in self-consciously creating a civil society, despite how their inappropriate nature for civil society. Bodin demanded that one respect the in-born goodness and reason of one's nature to realize an appropriate civil society (Bodin, *On Sovereignty*, p. 34, Par. 389). Grotius's contention that God himself is subject to the law of nature as humans are compelled to manifest it can be viewed as a transition in idea to the Modernist principle that speaks of natural law without attribution to God (Grotius, *The Rights of War and Peace*, p. 22).

Hobbes will write a Preface to *De Cive* after its publication a few years later, about 1647 or 1648 that fully cuts the cord to a fourth phasal articulation of the individual and society in conceiving society separate from a determined human nature, rather only a better invention through human thought and will. It will foresee the first paragraph of his *Leviathan* (1651). This becomes for all Western thought the major statement of individual choice in the construction of a society with the model of human proportion and agency recognized and clarified so that society becomes possible as a problem formulation and a problem solution. First, the first paragraph of his *Leviathan* of 1651:

> Nature (the Art whereby God hath made and governes the World) is by the *Art* of man, as in many other things so in this also imitated, that it can make an Artificial Animal. For seeing life is but a motion of Limbs, he beginning whereof is in some principall part within; why may we not say, that all *Automata* (Engines that move themselves by springs an wheeles as doth a watch) have an atificiall life? For what is the *Heart*, but a *Spring*; and the Nerves, but so many *Strings*; and the *Joynts*, but so many *Wheeles*, giving motion to the whole Body, such as was intended by the Artificer? *Art* goes yet further, imitating that Rationall and most excellent worke of Nature, *Man*. For by Art is created that great **Leviathan** called a **Common-Wealth** or **State,** (in Latin **Civitas**) which is but an Artificiall Man; though of greater stature and strength than the Naturall, for whose protection and defence it was intended; and, in which the *Sovereignty* is an Artificiall *Soul*, as giving life and motion to the whole body; The *Magistrates*, and other *Officers* of Judicature and Execution, artificiall *Joynts*; *Reward* and *Punishment* (by which fastened to the seate of the Soveraignty, every joynt and member is moved to performe his duty) are the *Nerves*, that do the same in the body Naturall; The *Wealth* and *Riches* of all the particular members, are the *Strength*; *Salus Populi* (the people's safety) its *Businesse*; *Counsellors,* by whome all things needful for it to know, are suggested unto it, are the *Memory*; *Equity* and *Lawes*, an artificiall *Reason* and *Will*; *Concord, Health*; *Sedition, Sicknesse*; and *Civil war, Death*. Lastly the *Pacts* and *Covenants*, by which the parts of this Body Politique were at first made, set together, and united, resemble that *Fiat*, or *Let us make man*, pronounced by God in the Creation.[7]

---

7. Thomas Hobbes, *Hobbes's Leviathan*, Reprinted from the Edition of 1651 (London: Oxford University Press, 1967), 8.

Hobbes' Preface to *De Cive (On the Citizen)* written after the 1641 edition, close to his 1651 *Leviathan*, begins with several key organizing concepts, much like the axioms we will see for first phases of all four metaparadigms we will track. His first paragraph of the Preface also contains the concept "method of reading," which again raises the significance of "being aware of how one is aware," and the significance of "method" in Modernist inquiry. Hobbes writes:

> I promise you, Readers, all that is usually thought to encourage attentive Reading: an important and useful Subject, a correct Method in the treatment of it, a good reason and an honest purpose in writing and good sense in the writer; and in this Preface I offer you a brief view of it all. This book sets out men's duties first as men, then as citizens and lastly as Christians. These duties constitute the elements of the law of nature, and of nations, the origin and force of justice, and the essence of the Christian Religion (so far as the limits of my design allow). (Hobbes, *On the Citizen*, p. 7)

Then, in the Ninth section of this Preface, he outlines in a cogent paragraph that covers his "method" of inquiry into the issues of men as citizens the self-wrought mechanics he amplifies in the first paragraph of the *Leviathan* of what he will call in the *Leviathan* the potential "art" of improving on nature:

> As far as my Method is concerned, I decided that the conventional structure of a rhetorical discourse, though clear, would not suffice by itself. Rather I should begin with the matter of which a commonwealth is made and go on to how it comes into being and the form it takes, and to the first origin of justice. For a thing is best known by its constituents. As an automatic Clock or other fairly complex device, one cannot get to know the function of each part and wheel unless one takes it apart, and examines separately the material, shape and motion of the parts, so in investigating the right of a commonwealth and duties of its citizens, there is a need, not indeed to take the commonwealth apart, but to view it as taken apart, i.e. to understand correctly what human nature is like, and in what features it is suitable and in what unsuitable to construct a commonwealth, and how men who want to grow together must be connected. (Ibid., p. 10)

This "taking apart" so that one as a reader can see the best or worst functioning of what has been "artificially" constructed as a society is the modus of the first phase in its initial concern with concepts and the structures they identify. A new final cause requires more than a name; it requires concepts that are redirected. Thomas Kuhn speaks of the manner in which key concepts of the older metaparadigm are preserved with new meanings that shift attention to the meaning of the new metaparadigm to be developed: "New paradigms are conservative: they preserve most of the terminology, examples, and methods of previous views" (1962, 7, 141–42, 168). An example of this will be Hobbes' well-known use of the previous metaparadigm's concept of "sovereignty," articulated by Jean Bodin that grants this power of determining governance only to dynastic royalty. Hobbes gives liberty as a natural birthright to every individual, regardless of station, and allows that anyone can choose to exercise that liberty to support any form of sovereignty—even as he argues for dynastic sovereignty. It will be used in this way to give personal freedom and choice to each person, one of the hallmarks of the new epoch of Modernism (Hobbes, *Leviathan,* pp. 100–101 [Chapter Fourteen]).

The first phase now breathes in the thought of new definitions, careful etymologies and redirection of older concepts. "Rhetoric" as Hobbes says is inappropriate for a first or second phase of an emergent metaparadigm that carries a new final cause—in Hobbes and the Modernism he helps found that is the fact all humans are the same in thought and competence—to speak with the rhetorical fluency of an argument. That will occur in the fourth phase, primarily, or, in a lesser expansive manner in the third phase of conflict and self-defense. In the first, and the second phases of the new final and the new formal cause, the elements of what will be constructed in reality must be examined.

Hobbes gives us not only the elements of a commonwealth that must be considered in the *Leviathan,* but an epistemology of human thought, part by part, and how ideation and emotion interact. Yet it is more of a short attempt at an encyclopedic presentation, much as what Diderot was to present a century later. One of the most thoughtful sections, which will be useful for his second phase of thought is Part I, Chapter Six, where develops definitions for states-of-mind in the individual to promote a proto-psychology of consciousness that will undergird "being aware of how one is aware." An example of only a few definitions shows the relevance even today of such an approach:

> For *Appetite* with an opinion of attaining, is called **Hope.**
>
> The same without such opinion, **Despaire.**
>
> *Aversion,* with opinion of *Hurt* from the object, **Feare.**
>
> The same, with hope of avoiding that Hurt by resistance, **Courage.**
>
> Sudden *Courage,* **Anger**. (Hobbes, *Leviathan,* p. 43 [Chapter Six])

The importance of individual freedom and autonomous choice which are perhaps the most significant principles of Modernism implicit or explicit in all fields of the arts and the sciences are testified to in a most striking manner by Hobbes in the *Leviathan.* He writes in the 21st chapter of Part II: "But if a man be held in prison, or bonds, or is not trusted with the libertie of his bodie; he cannot be understood to be bound by Covenant to subjection; and therefore may, if he can, make his escape by any means whatsoever" (p. 170).

Hobbes' definitional work as the basis of what he calls the rational law of nature had been begun in two works—*Human Nature* and *De Corpore Politico* (On the Political Body) written in 1640, but released to the public without his permission in 1650. These two works are essentially short definitional paragraphs of states mind in emotion, thought and kinds of action within the society. I suspect that writing them in the fourth phase of the Early Modern metaparadigm that ended c.1648, he felt that they were but incomplete terms that he was making clear to himself. Yet, by 1650 these definitional exercises into the character of human nature and political action were quite appropriate. I believe that having been born in 1588, his experience with the Early Modern notions of "being aware of being aware," and the freedom of choice of each individual had matured in his thought by 1640, when he was 52 years of age. The only difference from the Early

Modern will be his emphasis upon the autonomous individual, which was counter to the notions of Divine cause current in his society until the Modernist post-1648 vision.

One of the more interesting entries in *De Corpore* is where he speaks of the ancient Greek Riddle of the Theseus that demonstrates there is no necessary univocal objectivity, as answering the riddle of which is the real ship demonstrates various logically acceptable opinions. Having used the ancient ship riddle, Hobbes then sums up his realizations of the diversity and value of differing judgments:

> … whether a man grow old be the same man he was whilst he was young, or another man; or whether a city be in different ages the same, or another city. Some place *individuity* in the unity of *matter*, others in the unity of *form*, and one says it consists in the unity of the *aggregate of all accidents together*… and from hence springs a great controversy among philosophers.[8]

This testament to differing valid judgments of reality will be echoed by both Leibniz and Chladenius in the First Metaparadigm of Modernism that they shared with Hobbes, but it will disappear as univocal objectivity becomes a standard for all thinkers. While Hobbes and Leibniz will express the truth of multiple objective judgments of the same state-of-affairs, both men will overrule it in the standard definitions that they generate in their treatment of concepts. Leibniz famously said with his doctrine of "the identity of indiscernibles" that one could simply find a general consensus for such differences.[9] Modernism becomes the outlook that there is only one finding for sound inquiry. Postmodernism begins to dispute that.

There is another philosopher of history and historian who was contemporary to Hobbes, one whose focus upon taking apart institutional realities conceptually with well-conceived redefinitions, or newly coined definitions, exemplifies this first phasal focus of preparing a new final cause in human liberty and inquiry, separate from the God-given natural law that preceded post-1648 thinking. Samuel Pufendorf (1632–1694) wrote, in the spirit of "being aware of how one was aware":

> (political decisions) were influenced by the interaction of he "permanent" factors of geography, resources, and popular disposition, and the "transitory" factors of the changing situation and strength of neighbors, particularly as measured by the actual authority and policies of their rulers.[10]

---

8. Thomas Hobbes, *The Metaphysical System of Hobbes in twelve chapters from elements of philosophy concerning body*, ed. Mary Whiton Calkins, 2nd ed. (LaSalle, Il.: Open Court, 1963), 84. See also Mark E. Blum, "Continuity and Discontinuity, Change and Duration: Hobbes' Riddle of the Theseus and the Diversity of Historical Logics," in *Theory and Research in Social Education*, Vol. 24, 4 (Fall 1996): 360–90.
9. See Gottfried Wilhelm Leibniz, *Philosophical Letters and Papers*, ed. Leroy E. Loemker, 2nd ed. (Dordrecht: D. Reidel Publishing Company, 1969), 308, 505, 687, 692 and 699–701.
10. See Leonard Krieger, *The Politics of Discretion: Pufendorf and the Acceptance of Natural Law* (Chicago: University of Chicago Press, 1965), 185–86.

Pufendorf wrote a history of the German constitution, i.e., that of the Holy Roman Empire, in 1667, following his definitional criteria. As a product of the first phasal focus upon definitions and hypotheses the work did not give a systematic history of the Germanies, rather how its institutional structures were to be understood.[11] However, in 1682 he was to write a more narrative history of the great nations of Europe, following what I argue is the template for second phasal work in a metaparadigm—a systematic development of key causal concepts, i.e., the permanent and transitory factors of political decision-making, and their chronological occurrence that will guide more empirical inquiry in the third phase. I will review this second phasal work, *An Introduction to the History of the Principal Kingdoms and States of Europe* in the next chapter.

Hobbes' definitional work in philosophy of history can be linked to an actual history written shortly before the second phase began. Titled *Behemoth,* it can be considered a brilliant effort in systematically collocating his concepts on human thought, emotion and the intentions that one reads of in the definitional chapters of the *Leviathan* (1651), and showing them in operation in the motives and behaviors of historical personages. While *Behemoth* still displays a certain definitional character, it examines a precise, sequential period of years, 1640–1660, the time of the English Civil War.

Historical writing since Pufendorf and Hobbes in their post-1648 "life world" removed origins of the present from myth, informed by their Greek predecessors Herodotus and particularly Thucydides. When "final cause" was considered in the work of Pufendorf and Hobbes, it was to explain how the present had emerged from an evidentially accessible past.

In this first phase, Hobbes in his definitions ascribes final cause to a human nature that God is not referenced by, which is the need for "power" and its necessary expression as the core energy of a life. One's *Energeia* must be exercised would any *Ergon* be realized, that is a certain quality of life, form of "life world," or ethical or moral standard. Hobbes writes of this in Part I, Chapter X of the *Leviathan* (pp. 66–74). Sigmund Freud will make this understanding the foundation of his epistemology of human praxis. All history is the conflict of realizing one's sense of self in the "life world." History is the consequences over time of this necessary exercise of power in a world of others. A modernist history must show what, how, why, where and when over time this power is exercised and its consequences.

Hobbes was engaged in thinking about the narratives of human emotion and intention since his youth. He was a young man in his late teens and early twenties in the time of the great literature of Early Modern England. He is said to have been a friend of Ben Jonson, and had known Shakespeare. There is an interesting study of the relationship of the proto-psychological insights of *Leviathan* and Shakespeare's *Macbeth*.[12] I believe this

---

11. See Samuel Pufendorf, *The Present State of Germany,* trans. Edmund Bohum, 1696, edited with an Introduction by Michael J. Seidler (Indianapolis: Liberty Fund, 2007).
12. Nicholas Dungey, "Shakespeare and Hobbes: *Macbeth* and the Fragility of the Political Order," SAGE Open April-June 2012: 1–18 © The Author(s) 2012 DOI: 10.1177/2158244012439557 http://sgo.sagepub.com.

interest carried through his entire life, but in his thought during the fourth phase of the Early Modern metaparadigm, he hesitated to have his definitional exercises and proto-psychology published; which I see as the realization that it did not conform to the collective narrative norms of explanation of that time. Thus, his definitional exercises in *Human Nature* (1640), and *De Corpore* (1640), were withheld by him from publication, and only published in 1650 without his knowledge, such definitional knowledge at that time seen as not yet voiced in an emergent individualistic age.

B. Literature—The Modern Novel

Anton Ulrich (1633–1714)

If one considers the core of the Western Modern novel, in our sense of post-1648 thought, it can be dated to writing from the earliest classical age of Greece, in the thoughtful dialogues and consequent actions between the protagonists of Aeschylus, Sophocles and Euripides. Greek drama does contain as its core the fundamental principle of an individualism that is self-directed by choice, and the principle of "be aware of how one is aware," which makes the Modern novel into a development of personal character with changes in awareness that are central to the story. One may find this self-reflectiveness in the drama of Euripides' *Iphigenia in Tauris*, but as all such insights in Western thought as depicted in fiction, through Dante, into Boccaccio, and Rabelais, were limited and essentially non-developmental in its focus upon character. Modernism instills changing self-temporality as a sense of self-development into the durational sense of years. That is why modern history is finally written stressing a complex view of personal agency and the "life world" of secular influence. Cervantes' writings, such as *The Man of Glass* is a genuine precursor in the sense of the protagonist "being aware of his mode of awareness, and those of others," lacking only the element of the narrative development over time. His *Don Quixote* arguably does have development of time in insight and personal character between Parts I and II. I will call that "genius" with the kind of foresight that I believe Thomas Hobbes was developing in the fourth phase of the Early Modern as he refined his definitions of rational and emotive intentions. Nonetheless, I will begin the truly modern novel that is forthright in its character development with the work of Anton Ulrich and his novel *Aramena* (1669). In the final years of the 1660s one begins to see the second phasal development of a formal cause primary in thought in the various disciplines, and Aramena has by that time fully developed a systematic plot of change over time for its characters. Yet, I classify the novel in its core characteristics of Modernism from the first phase when its ideation and composition began, dated by its most careful scholar, Blake Lee Spahr, as possibly before 1663.[13] There was a "modern" novel before this arguably, but not one that has the characteristic treatment of persona and action that I will contend makes *Aramena* the first. La Calprenéde's *Cléopatre* was written between

---

13. Blake Lee Spahr, *Anton Ulrich and Aramena: The Genesis and Development of the Baroque Novel* (Berkeley and Los Angeles: University of California Press, 1966), 11, 55–59.

1647 and 1658. Anton Ulrich is said to have read it on a visit to Paris in 1656,[14] but his use of that text is more informative of the romance tradition than that of the modernist novel form I will describe.

Blake Lee Spahr in an astute article speaks of the significance of the novel for helping its society of readers deal with the challenge of "becoming aware of one's own form of awareness." The novel's focus upon the intentions of everyday life, developing common-sense insights into how one experienced life and affected others, eased the traumatic influence separation from the Divine in an everyday sense. The final cause of the Divine was shifting toward what Suarez called "proximate cause" and Hobbes pragmatically substituted "life's primary responsibility is that of choice". Hobbes might make this separation, but ordinary people were compelled to live it, as the "separation of church and state." Thus, the birth of the novel as a transition for individuals to the new final, formal, material and efficient causes of their "life world." Spahr, quotes Angelus Silesius (1624–1677), a militant Catholic priest and essayist, saying "I do not know what I am, I am not what I know."[15] What he and others of that time did to find an identity that was functional, even if not answering the issue of God's presence, was to examine their social roles in public and in private. The writers of that time, generating what became with Anton Ulrich the first Modern novel, transformed the previous "romance" style of the "Schäferroman" (the "pastoral romance"), which featured encoded real person-alities behind the masks of the characters of the plot, into the idea of changing social masks given the type of intercourse among the individuals of their "life worlds." In the "pastoral romance" the reader was challenged to discern through the behaviors of the persona of the tale living individuals in his or her environment by the character of their language and actions (Spahr 1968, 18–19). Thus, a "becoming aware how others were aware" was the challenge of the literature, roughly between 1630 and 1660. The use of that skill now became a part of the character in the new novel, and, an expectation of the same insight in the reader (Spahr 1968, 19). This was the seminal beginning of soci-ology as focused upon how an individual with self-awareness—being aware of how one is aware—altered his or her behavior in language, posture and ideation, appropriate to the situation. This was not a new reality, but a swimming with the now emphasized secular "life world." God remained the higher reality, but one could not be certain either of his presence or when and how one conformed to the Divine intent that had been the philos-ophy of predestined individuation. Thus, Angelus Silesius' statement. Indeed, Pascal in 1656 spoke of his "wager" that God existed,[16] and even Hobbes in the late 1670s spoke

---

14. See Anna M. Schnelle, *Die Staatsauffassung in Anton Ulrich's "Aramena" im Hinblick auf La Calprenédes "Cléopatre."* Inaugural Dissertation zur Erlangung des Doktrgrades genehmigt von der Philosophischen Facultät der Friedrich-Wilhelm-Universität zu Berlin (Berlin: Verlag Rudolph Pfau, 1939), 84.

15. Blake Lee Spahr, "Der Barockroman als Wirklichkeit und Illusion," in *Deutsche Romantheorien, Beiträge zu einer historischen Poetik des Romans in Deutschland*, hrsg. Reinhold Grimm (Frankfurt am Main: Athenäum Verlag, 1968), 19–20.

16. Blaise Pascal, *Pensées and the Provincial Letters* trans. Thomas M'Crie (New York: The Modern Library, 1941), 79–84 [*Pensées*, Par. 233].

of his final voyage, "a leap into the dark,"[17] trusting that the separation between God and human intention might be healed after death.

The structure of *Aramena* is indefinite in its sequence of years, something that will be corrected when Anton Ulrich writes his second novel, *Die Römische Octavia* in the second phase of the first Modernist metaparadigm, c.1677–1679. *Aramena* takes place during the patriarchate of the Biblical Jacob in the countries of the Near East. *Die Römische Octavia* takes place over a two-year period from 68 AD to 70 AD, the end of the Empire under Nero until the first year of Vespasian. The text of *Die Römische Octavia* is careful to attend to historical sources, whereas the fictional indefiniteness of *Aramena* serves as poetic license to experiment with the issue of social roles, masks and the demand of discerning how social personas change in relation to the societal milieus. In this way, Anton Ulrich's *Aramena* is analogous to Hobbes' *Leviathan,* and his earlier *De Cive* and *De Corpore,* that is, it is an exercise in definitions more than a systematic, developmental sequence of concepts that can be shown to occur sequentially over time, as the second phase of every metaparadigm calls for. *Die Römische Octavia,* on the other hand, details the self-development of Octavio over time as she comes to be aware of her "knowing when, how, and why" she becomes aware of certain states of awareness demanded of her in her social interactions, and her growth of character with that knowledge.

Aramena is distinctive for having 36 episodes that present the interactions of 300 persons whose individual experiences each have sufficient depths to characterize ways of self-presentation with the differing normative standards for interaction in the particular milieu presented (Schelle 1939, 6–7). To further deepen the necessity of a careful reading toward the complexity of individual characters, there are more than four of the 300 who have the same names (ibid., 6). The reason for this structure I contend is that Anton von Ulrich is practicing the short conceptual sketches Hobbes practices to clarify definitions of thought in its concepts, emotional affects and other aspects of social intercourse. Anton von Ulrich compels the reader to become aware of how one characterizes their social masks to conform to dialogical situations. The destiny of one or two persons is not as important as the coming to know how one becomes aware oneself in life situations where choices of self-presentation and insight into the other with whom interacts is primary. Here are several examples of minor characters in their dialogue with others:

> I will say it again, answered Corycide, being a shepherd in the most fortunate thing in the world. Those people who simply live on the land as myself do not recognize how fortunate they are in comparison to those who live at court, because they have not experienced the bitterness of the life there. I know better as I have lived at both places.[18]

Corycide speaks for himself in answer to Ephron, who has begun the comparison to the two lifestyles and their affect on those who live one or the other. The personalization of the speaker and the direct use of verbs and prepositions and adverbial phrases that offer

---

17. https://nottinghamcityofliterature.com/blog/a-great-leap-in-the-dark-thomas-hobbes
18. Anton Ulrich, *Aramena*, ed. Dieter Merzbacher (Stuttgart: Anton Hieremann, 2017), 3.

the facts of how one's own self-understanding is affected by one topos or another are repeated throughout the conversations in the text.

Ahalibama shortly later tells her story, one of the main 36 episodes. One passage from that story again shows first person narration based upon reflection from memory that takes apart the past situation in its possible ways that she might have acted given the normative understandings of such situations:

> I did not know if I should have taken the gift from him or his servants who had delivered it for him, but this occurred an hour before I awoke. Timna took it in my name. I was vexed about her doing this when I awoke, having slept too long. I reflected upon the cultivated conversation in the first part of the evening, during which Timna perhaps spoke the truth that in my weakness I did not share with her the truth of matters, and I became depressed that she had noticed this. So, I now vaunted my rage upon someone, Timna, who was not guilty, Timna now pointing out my state-of-mind, asked me not to infect him when he came with my vexed attitude, as I looked at him. I then thanked him when I saw him, while not forgetting that I told him the night before I would not be an ostrich, and not give him my permission (in his requests). He answered me then somewhat saddened that I had punished him sufficiently for impropriety. (Ibid., 19)

Ahalibama goes on for a page letting us—the readers—see her shifting thoughts, recalling states-of-mind, thinking of what is appropriate in behavior toward her situations with Elieser, while not mentioning his name. This referring only to "he" becomes a project testing the reader's discernment of who spoke to whom, thus creating an intensive set of reflections in the reader on the behavior and self-justifications presented. Elieser had been introduced by name once two pages before. The reader must carefully go back and forth to see the past experiences with him and others around him before he is finally mentioned by name once again two pages since his first such mention. We see complex shifts in attitude in this durational event between him and Ahalibama that even today make sense in terms of how we enter situations given past experience with another and cope with the present. We are informed by how one presents oneself to others given x, y or z past encounters. Anton Ulrich gives us instructions that 400 years later still resonate. This is the first phase of the first Modernist metaparadigm teaching us how "to be aware of how we are aware."

C. Mathematics—The Development of Logical Calculus (A Precursor of Calculus)

Gottfried Wilhelm Leibniz (1646–1717)

Leroy Loemker writes of Leibniz's development of a logical calculus in 1666:

> The work contains the germ of the plan for a universal characteristic and logical calculus, which was to occupy his thinking for the rest of his life. That project is here conceived as a problem in arithmetical combination of simple into complex concepts. Leibniz deriving basic theorems on permutation and combination and applying them to the classification of cases

in logic, law, theology, and other fields of thought. His later judgment on the work was that in spite of its immaturity and defects, especially in mathematics, its basic purpose was sound. [19]

The "logical calculus" led through refinements in modernist thought to what now is called "symbolic logic." Two of the most authoritative mathematicians in the twentieth century, Hans Reichenbach and Rudolf Carnap, refer to Leibniz in their main texts.[20] Leibniz's logical calculus was and is central to my own field of historical hermeneutics in analyzing the structure of sentential judgments.[21]

Leibniz's "logical calculus" was meant as a way of creating a definitional system for being able to demonstrate the logical connection of the parts of propositions—how they are linked causally, and their implications, in a manner that precedes empirical verification, but enables thought to embark on focus inquiry. As with Hobbes or Anton Ulrich, we see the definitional emphasis of thought in this seminal, first phase of the new metaparadigm, which will be Modernism.

Leibniz writes, for example, in his consideration of a proof of God, the following sequence of propositional parts and their relations, using the mathematical symbols of a logical calculus—A, B and C:

Hypotheses [Praecognito]

1. Definition 1. *God* is an incorporeal substance of infinite power [virtus]/
2. Definition 2. I call *substance* whatever moves or is moved.
3. Definition 3. *Infinite power* is an original capacity [*potential*] to move the infinite. For power is the same as original capacity; hence we say that secondary causes operate by virtue [*virtus*] of the primary.
4. Postulate. Any number of things whatever may be taken simultaneously and yet be treated as one whole. If anyone makes bold to deny this, I will prove it. The concept of *parts* is this: given a plurality of beings all of which are understood to have something in common; then, since it is inconvenient or impossible to enumerate all of them every time, one name is thought of which takes place of all the parts in our reasoning, to make the expression shorter. This is called the *whole*. But in any number of given things whatever, even infinite, we can understand what is true of all, since we can enumerate them all individually, at least in an infinite time. It is therefore permissible to use one name in our reasoning in place of all, and this will in itself be a *whole*.

---

19. Gottfried Wilhelm Leibniz, *Philosophical Papers and Letters: A Selection Translated and Edited with an Introduction by Leroy E. Loemker*, 2nd edition (Dordrecht: D. Reidel, 1969), 73–74.
20. See Hans Reichenbach, *Elements of Symbolic Logic* (New York: Macmillan, 1948), 210, and Rudolf Carnap, *Introduction to Symbolic Logic and Its Applications*, trans. William H. Meyer and John Wilkinson (New York: Dover, 1958), 17.
21. See Mark E. Blum, *Continuity, Quantum, Continuum, and Dialectic: The Foundational Logics of Western Historical Thinking* (New York: Peter Lang, 2006), especially, 113–46. Also, Mark E. Blum, *Cognition and Temporality: The Genesis of Historical Thought in Perception and Reasoning* (New York: Peter Lang, 2019), especially 25–58.

5.   Axiom 1. If anything is moved, there is a mover.
6.   Axiom 2. Every moving body is being moved.
7.   Axiom 3. If all its parts are moved, the whole is moved.
8.   Axiom 4. Every body whatsoever has an infinite number of parts; or, as is commonly said, the continuum is infinitely divisible.

[Here Leibniz shows his later stress on the continuum in his development of differential and integral calculus. M.E.B.]

9.   Observation: There is a moving body.

Proof:

1.   Body is *A* in motion, by hypothesis No. 9.
2.   Therefore there is something which moves it, by No. 5.
3.   and this is either incorporeal
4.   because it is of infinite power, by No. 3;
5.   since *A*, which it moves, has infinite part, by No. 8;
6.   and is a substance, by No. 2.
7.   It is therefore God, by No. 1 Q.E..D.
8.   Or it is a body,
9.   which we may call *B*.
10.   This is also moved, by No. 6,
11.   And what we have demonstrated about body *A* again applies, so that
12.   Either we must sometime arrive at an incorporeal power, as we showed in case of *A*, in steps 1–7 of the proof, and therefore God;
13.   or in the infinite whole there exists bodies which move each other continuously.
14.   All these are one whole that can be called *C* by No. 4.
15.   And since all the parts of *C* are moved, by step 13,
16.   *C* itself is moved, by No. 7,
17.   and by some other being, by No. 5,
18.   namely, by an incorporeal being, since we have already included all bodies, back to infinity, in *C*, by step 14. But we need something other than *C*, by 17 and 19,
19.   which must have infinite power, by step No. 3, since C, which is moved by it, is infinite, by steps 13 and 14;
20.   and which is a substance, by No. 2,
21.   and therefore God, by No. 1.

Therefore, God exists. Q.E.D.

The logical calculus of A, B and C will become a more complex mathematical helpmeet in demonstrating propositions by the second phase of this first Modern metaparadigm. It will become a systematic theory of proving sentential meaning in greater detail as propositions regarding any subject—be it theology, history or mathematical analysis itself is concerned. The more systematic logical calculus of Leibniz as it is affected by the structural demands of the second phase of any metaparadigm will be demonstrated in the next chapter. The second phase of any metaparadigm demands a systematic development of the guiding concepts in its new problem-solving orientation. Then, the more intensive empirical work into thought or material nature offered by the third phase of any metaparadigm can be carried out in accord with its system.

Central to both the logical calculus and the subsequent development of differential and integral calculus was how the initial movement of an A was caused, how sustained and how explained in its direction and possible acceleration. The idea of movement becomes central both to Early Modernism and the Modernism that follows. "Change over time" is not only the foundation for the writing of modern history, it is the foundational concept for any study of inorganic, organic or human nature. The significance of problem solving, in itself, accounts for this concern. Leibniz, and as we will discuss, Isaac Newton were deeply concerned throughout their lives with this problem of movement and changing velocity mathematically, generating differential and integral calculus to solve the contained issues. Hobbes, too, was fascinated by the issues, stimulated by Galileo's conception of the laws of motion, in terms of inertia and changing velocity.[22] A key concept all thinkers engaged in these concerns had to address was the concept of "conatus." Conatus is the initial movement of an X. What creates the initial movement and sustains it? Is it from within, a secondary final, spiritual cause as well as a "proximate" efficient cause in the sense that Suarez and Leibniz were to view it and measurable mathematically? Or, as with Newton, was there an internality, non-spiritual, but a force that impelled a growing magnitude of force that surrounded and affected the X in its inertial movement, demanding a complex new system of measurement, and thus possible predicative understandings. Or, was it as Hobbes viewed conatus, an external, material cause, carried further by the inertial velocity of the material mover, but harboring none of the internal complexities of force that called for measurement? Hobbes' pragmatic, but mathematically "naïve" understanding (Boyer 1949, 178–79) was one of the major addresses of Aristotle's four causes that separated human science from God's intervention in the secular norms of thought that became normative. Yet, this pragmatic view of conatus as a causal concept hindered his taking up the necessary inquiry into the complexity of the problems of sustained movement, changing velocity and direction that led Leibniz and Newton to their development of differential and integral calculus.

Carl B. Boyer in discussing the difference in the approaches of Newton and Leibniz to this problem of conative beginnings articulates an interesting difference in the thought of Germans and English thinkers that mark all inquiry in all fields in the age of Modernism:

> Whereas Newton used the physical idea of the (conative) "moment" of growing magnitudes, there grew up in Germany a more metaphysical form of this in the notion of *intensive* magnitude as opposed to *extensive* quantity. Upon this idea of a "tendency" or a "becoming" the mathematical infinitesimal throve, with the result that philosophers have been reluctant to abandon it, even though modern mathematics has shown that the basis of calculus is to be found in the derivative rather than in the differential. (Boyer 1948, 178–79)

---

22. See Carl B. Boyer, *The History of the Calculus and Its Conceptual Development* (New York: Dover, 1949), 177–78.

Since Martin Luther and the idea of predestination, this spiritual, and later, genetic stress on internal force as the conative origin of human movement has persisted in all the disciplines of the arts and sciences in Germany. This societal difference among these nations will be significant to understand, even as all Western nations shared the same metaparadigmatic understandings in their temporal activities.[23]

---

23. I discuss these societal differences in the nations of the United States, France, England, and book length in Germany and Austria in Mark E. Blum, *German and Austrian-German Historical Thought in the Modern Era* (New York: Lexington Books, 2020), especially vii–xxxv.

Chapter Two

THE SECOND PHASE: DEVELOPING A SYSTEMATIC THEORY FOR FUTURE INQUIRY AND PROBLEM-SOLVING C.1670–C.1690

The real interest (of a nation) may be subdivided into a perpetual and a temporary. The former depends chiefly on the Situation and Constitution of the country, and the natural Inclinations of the people; the latter on the Condition, Strength and Weakness of the neighbouring nations; for as those vary, the Interest must also vary. Whence it often happens, that whereas we are, for our own Security sometimes oblig'd to assist a neighbouring Nation, which is likely to be oppress'd by a more potent Enemy. We are at another time for'cd to oppose the Designs of those we before assisted, when we find they have recover'd themselves to that degree, as that they may prove Formidable and Troublesome to us.

But seeing this Interest is so manifest to those who are vers'd in the State-Affairs, that they can't be ignorant of it, one might ask, How it often times happens that great Errors are committed in this kind against the Interest of the State. To this may be answer'd that those who have the Supreme Administration of Affairs are oftentimes not sufficiently instructed concerning the Interest both of their own State, as also that of their Neighbours. And yet being fond of their own Sentiments, will not follow the Advice of understanding and faithful Ministers. Sometimes they are misguided by their Passions, or by Time-serving Ministers and Favourites. But where the Administration of government is committed to the Care of Ministers of State, it may happen that these were not capable of discerning it, or else are led away by a private Interest which is opposite to that of the State. Or, they may be divided into Factions, being more concern'd to ruin their Rivals than to follow the Dictates of Reason. Therefore some of the most exquisite parts of Modern History consists in that one knows the Capacity, Inclinations, Caprices, Private Interests, manner of proceeding, and of (those responsible for policy), since upon this depends, in a great measure, the good and ill management of a State. For it frequently happens that a State, which in itself is consider'd is but weak is made to become very considerable boy the good Conduct and Valour of its Governours. Whereas, a powerful State (can be made weak) by the ill management of those at the Helm, oftentimes suffering considerably. But as the Knowledge of these Matters appertains properly to those who are employ'd in the management of Foreign Affairs, so it is mutable, considering how often the Scene is chang'd at Court. Wherefore it is better learn'd from Experience and the Conversations of Men well vers'd in these Matters, than from any

Books whatsoever. And this is what I thought my self oblig'd to touch upon in a few Words in this Preface.[1]

Samuel Pufendorf, *An Introduction to the History of the Principal Kingdoms and States of Europe* (1682)

Pufendorf in this Preface to a work that will briefly be reviewed below in the spirit of what is the second phasal thought of any metaparadigm, that is a systematic set of concepts which can guide more in-depth research, indeed argumentative contesting of the findings of other historians. The leading causal consideration of any second phasal work in any discipline is that of Aristotelian formal cause. That is, the sequential pattern of concepts and axioms deemed necessary for a thorough problem solving, these concepts and axioms being introduced and explained in the first phase of the metaparadigm. Pufendorf in this 1682 text will review both the perpetual and temporary factors of a nation's historical chronology and development, its changes and either loss or increase of power within and among other nations. We will see this systematic development in the service of ongoing problem solving in each discipline of the metaparadigm.

A. History/Philosophy of History

Thomas Hobbes (1588–1670)

In time, as in place, there were changes high and low. I verily believe that the highest time would be that which passed between the years of 1640 and 1660. For he that thence, as from the Devil's Mountain, should have looked upon the world and observed the actions of men, especially in England, might have had a prospect of all kinds of injustice, and of all kinds of folly, that the world could afford, and how they were produced by their "dams" hypocrisy and self-conceit, whereas the one is double iniquity, and the other double folly.[2]

The Behemoth is a favored creature of God's creation (Job, 40: 15–24), whereas the Leviathan is evil, although as indomitable (Isaiah, 27: 1; Job, 41: 1–34)). One could argue that Hobbes saw value in the Leviathan (Job, 41: 32–34). Hobbes would have each person with an ultimate authority, and a freedom that brooked no law or covenant (*Leviathan* 1967, 170 [Art 2, Chap. 21]). Whereas Hobbes' *Leviathan* takes apart the commonwealth and human nature part by part, the *Behemoth* is a historical text that argues how it can be re-constituted, as a better solution than that which occasioned and endured an actual civil war. The *Behemoth* as a second phase historical project is a systematic philosophy of political science that spells out the guidelines for the actual re-construction of England after Hobbes dating of its Civil War from 1640–1660, the final decade being for Hobbes the unlawful, evil governance of Cromwell and his son.

---

1. Samuel Pufendorf, *An Introduction to the History of the Principal Kingdoms and States of Europe*, trans. Jodocus Crull (1695), ed. Michael J. Seidler (Indianapolis: Liberty Fund, 2013), 8–9.
2. Thomas Hobbes, *Dialogue, Behemoth, & Rhetoric* (Whithorn, England: Anodos Books, 2019), 73 [*Behemoth*, Dialogue I].

The *Behemoth* is written as a dialogue between A and B. This gives it a narrative character of ordinary language. All writings in the humanities and social sciences will take on that character in the epoch of Modernism, as it privileges the in-common of human nature and the individual in the ordinary discourse that creates this leveling of understanding. To be sure, technical works are written beyond a common narrative language, but astute thinkers know they are contributing to this new epoch of every person, especially since Hobbes locates sovereignty in each person. A narrative of ordinary language ensures that it can serve to enlist inquirers to flesh out its systematic purview.

The pedagogic definitional address of clearly separated concepts does continue throughout the text, bringing forward the definitional work of his earlier *Leviathan*. Yet, the narrative has sufficient leaven of monumental events in chronological order to still be a history. Monumental events, the term to be understood in the sense Nietzsche writes of them in his 1874 *The Advantage and Disadvantage of History for Life*, where the selection creates a trajectory that accommodates and furthers the philosophy of occurrences the narrator holds and wishes to convince his reader with. Yet, it is an objective story, as the selection of events is sound in giving both sides of the Civil War an airing. Hobbes chooses pertinent legal or canonical concepts that address the issues within Charles I's reign, the rise of power in the Parliament, Parliament's justifications of authority when Charles I is on the run and how Cromwell justifies his own executive authority. As the historical narrative begins, Hobbes constructs a political-social-ideological context of conflicting groups—the tenor of Parliamentary opinion, the beliefs of diverse Protestant groups who oppose the King, in the main the Presbyterians, and the continuing efforts of the Pope, through English Catholics, to interfere in English government. The major ideas of authority, sovereignty and even the necessity of a strong military within a monarchical sovereignty, keep the narrative in touch with a lexicon of precise definitions, appropriate to the succession of events, even as Hobbes will favor that which favors monarchy.

An example of Hobbes' narrative style below typifies his integration of defined ideation, normative understanding among differing milieus, and Hobbes' own subtle bias in presentation. In this case, not only many of the King's ministers, but the very army that fought to protect Charles I in 1643 (Hobbes 2019, 73, 137) is shown to have a popularly filtered understanding of the ideas involved that work against their service to him:

> For who was there of them though knowing the King had the sovereign power, that knew the essential rights of sovereignty? They dreamt of a mixed power, of the King and two Houses. That it was a divided power, in which there could be no peace, was above their understanding. Therefore they were always urging the King to declarations and treaties (for fear of subjecting themselves to the King in an absolute obedience); which increased the hope and courage of the rebels, but did the King little good. For the people either understand not, or will not trouble themselves with controversies in writing, but rather, by his compliance and messages, go away with an opinion that the Parliament was likely to have a victory in the war. Besides, seeing the penners and contrivers of these papers were formerly members of the Parliament, and of another mind, and now revolted from the Parliament, because they could not bear the sway in the House which they expected, men were apt to think they believed not what they writ. (Behemoth, 143)

The next paragraph offers objective events, carrying the military narrative of the Civil War chronologically further:

> As for military actions (to begin at the head-quarters) Prince Rupert took Birmingham, a garrison of the Parliament's. In July after, the King's forces had a great victory over the Parliament's, near Devizes on Roundway-Down, where they took 2,000 prisoners, four brass pieces of ordinance, twenty-eight colours, and all their baggage; and shortly after, Bristol was surrendered to Prince Rupert for the King; and the King himself marching into the west, took from Parliament many other considerable places. But this good fortune was not a little allayed by his besieging of Gloucester, which, after it was reduced to the last gasp, was relieved by the Earl of Essex; whose army was before greatly wasted, but now suddenly recruited with the trained bands of apprentices of London. (Hobbes 2019, 143)

All the above is spoken by A, which we assume is Hobbes's instructive voice. As the entire historical narrative is a dialogue with B, B's comment to the above is instructive insofar as Hobbes' bias:

> B.   It seems, not only by this, but also by many examples in history, that there can hardly arise a long or dangerous rebellion, that has not some such overgrown city, with an army or two in its belly to foment it. (Hobbes 2019, 143)

Yet, this is after all a second phase product of a metaparadigm, and what Hobbes does with his interlocutor is to assure further research is possibly done by readers to verify this factual insight.

As the history continues, Hobbes seems to remind himself that a proper *historia* (in the sense of Herodotus, i.e., both an objective inquiry and a story) requires putting the reader into an environ as closely as possible, so it can be experienced. Indeed, this is the tactic of the new novel form in its alignment of interactions that can be seen as occurring in the reader's everyday life. Hobbes writes of Charles I's attempt to escape to France:

> The King, therefore, in a dark and rainy night, his guards being retired, as it was thought, on purpose, left Hampton Court and went to the sea-side about Southhampton, where a vessel had been bespoken to transport him, but failed; so that the King was forced to trust himself with Colonel Hammond, then governor of the Isle of Wight; expecting perhaps some kindness from him, for Dr. Hammond's sake, brother to the colonel and his Majesty's much favoured chaplain. (Hobbes 2019, 153)

"It was … a dark and rainy night …". Was this the inception of this image of a narrative beginning that is but comic relief today? What Hobbes attempts here is to bring the material, secular realities of nature that the individual must cope with.

Hobbes brings into this episode the kind of inquiry into complex intentions that his earlier definitional approach in *Leviathan* took up. He infers that the guards were instructed by Cromwell to allow Charles I to escape hoping he would flee England, as a solution for how to handle him at a time in the struggle too early to put him to death without repercussions from the populace:

(Cromwell)'s main end was to set himself in (Charles I)'s place. The restoring of the King was but a reserve against the Parliament, which being in his pocket, he had no more need of the King, who was now an impediment to him. To keep him in the army was a trouble; to let him fall into the hands of the Presbyterians had been a stop to his hopes; to murder him privately (besides the horror of the act) now whilst he was no more than lieutenant general, would have made him odious without furthering his design. There was nothing better for his purpose than to let him escape from Hampton Court (where he was too near the Parliament) whither he pleased beyond the sea. For though Cromwell had a great party in the Parliament House whilst they saw not his ambition to be their master, yet they would have been his enemies as soon as that had appeared.—To make the King attempt an escape, some of those that had him in custody, by Cromwell's direction told him that the adjutators meant to murder him; and withal caused a rumour of the same to be general spread, to the end it might that way also come to the King's ear, as it did. (Hobbes 2019, 153)

Samuel Pufendorf (1632–1694)

In order to show now the parallel movement in historical writing in the second phase of this first Modern metaparadigm, I will examine Samuel Pufendorf's 1682 *An Introduction to the History of the Principal Kingdoms and States of Europe*. In his 1667 primarily definitional work on the institutions of the Holy Roman Empire and the German states, he had developed two key conceptions that would further all later historical narratives, consciously or unconsciously in later historians. These were that the history of a society were conditioned by two factors that entered the policy decisions of its authorities, "permanent factors" that included geography, resources, and popular disposition of its members, and the "transitory" or "temporary" factors of the changing situation and strength of neighbors, particularly as measured by the actual authority and policies of their rulers. Pufendorf's 1682 history of the several nations dominant in Europe, including England, applied these categories to each narrative section that took up one country. An example of his systematic application of permanent and temporary factors is his overview of the history of England (or Britainy).

The island nature of the country is in the first paragraph, and, Pufendorf cites that fact as constraining any sustained policies of other nations insofar as a permanent invasion. Even the Romans could not conquer all of the island "the utmost parts of Britainy being almost inaccessible" (Pufendorf 2013, 116). Then, Pufendorf goes chronologically from the Saxon heptarchy from 455 AD through 724 AD. Then, Egbert, King of the West Saxons, subdued the others either with force or threat, and became the single King of England in 818 AD. The policies then of Egbert through Ethelred, who began his reign in the year 979 AD, in relation to the invading Danes are sketched. The policies of Egbert, Ethelred and his son Edmund toward the invading Danes are then offered in several sentences. One can see this as the permanent factor of the Danish success in their centuries-long invasions. A more fulsome history of this time is shown to be necessary, and the broad outline of the succession of geographical conquest of island is sketched. Personalities of the rulers and advisers are not given. Archival material may later inform the kind of depth of narration that will fill in the "transitory" or "temporal" issues of the policy makers (Pufendorf 2013, 118–19).

The Danish control ended in 1041 AD with the death of Hardiknut, who, it seems died from overeating, presumably a heart attack. This is the sole character attribution to this point in the history, Hardiknut doing nothing so far as policy except continuing to tax the English Saxons heavily. The English Saxons created a holiday, "hock-tide" to be celebrated the second Monday and Tuesday after Easter thenceforth, the death of Hardiknut an event that called forth their hatred and opposition to the Danes, who actually ruled only 26 years. This mention by Pufendorf gives us the sentiment of the people, and thus helps us understand the succeeding events, as this sentiment is shown to be a "permanent" factor of England's peoples. Yet, it is at the same time a reference to the character of the ruling authority insofar as significant policy (Pufendorf 2013, 119–20).

The reign of Edward the Confessor, an Anglo-Saxon of the House of Wessex followed in 1042. Edward remitted the tax of Danegeld, winning thereby the affection of the people (again a permanent factor of the sentiment of the population). When the Normans, under William the Conqueror, subdued the Saxons in 1066 the Anglo-Saxons were a strong, aristocratic as well as common population. The leadership of William, however, Pufendorf praises. Pufendorf writes: "The *English* were at first extreamly satisfy'd with his Government. He leaving each in possession of what was his own, only giving the vacant lands to the *Normans*; partly, also because he was related to the former Kings of *England* through Edward the Confessor, partly because he was recommended to them by the Pope" (Pufendorf 2013, 122). Again, here is a policy, particularly that of only taking vacant lands for the Normans who accompanied him, that explains a "transitory" factor that can be said to become sufficiently permanent to explain why and how the Magna Carta was achieved by the non-Norman aristocrats in 1215.

Surprisingly, Pufendorf never mentions the Magna Carta of 1215, his interest being in the succession of Kings and then Queens and their foreign policies toward neighboring countries. Pufendorf goes on to offer a few lines or a paragraph about the "transitory" policies that depend upon the decisions of the particular monarch, all the way up to Charles II (1660–1685), the present King as he wrote. He often gives an anecdote that is to clarify policy choices by the particular monarch, but an anecdote only. One of my favorites is that of Canute, the Dane who ruled as King of England from 1017 AD to 1035 AD:

> Some of his Parasites, who pretended to attribute to him something above a Humane Power, he ridicul'd, by causing a Chair to be set near the Sea-side, commanding the Seas not to wet his Feet; but the Tide rolling on the Waves as usually, he told them That from thence they might judge of what extent was the Power of all worldly Kings. (Pufendorf 2013, 119)

The individual appears in each anecdote he offers throughout the book with both human foibles and approachable natures that ring true today.

Pufendorf creates a Modernist model of monarchs who in general "are aware of being aware" and are able to laugh at themselves, as one must say of Canute. Pufendorf creates a pattern of the permanent and the transitory that will create a template, as did Hobbes's *Behemoth* for the deeper, more expansive inquiry and narrative that will follow.

B. Literature

Anton Ulrich (1633–1714)

One sees ancient Rome in its mouldering decay, yet out of these ruins a new time is evident. The new science is apparent in old wars, wherein a new maxim of statehood emerges from the felled pillars.

Tribute to the Author of *Die Römische Octavia* in praise of its
happy ending, by a contemporary of Anton Ulrich.[3]

This second novel, *Die Römische Octavia* by Anton Ulrich will use the same complex reflections by its characters, indicating their own "awareness of how they are aware" and the subsequent ability to change behavior based upon that, as in his first novel *Aramena*. The difference will be in the concentration upon the heroine and her later consort giving a longer trajectory to the one story, and by that creating a pattern for what will become a more thorough and detailed novel form by the third and fourth phase of this first Modern metaparadigm. Moreover, the years of the story will be shorter, based largely on historical events that can be inquired into, and thus generating besides its more complex individualities, the bases for literature to become a means of showing patterns of thought and consonant behaviors that were to become normative in the Modern age. When Hayden White in the later Modernist age we are now in emphasizes that fiction can teach us about history, Anton Ulrich's *Octavia* can be seen as at least its first secular, seminal model.[4]

The story takes place between 68 AD and 70 AD in the Roman Empire. This is the period in which the assassinations of Emperors Nero and Sulpicius Galba occur, the civil war between Oto and Vitellisu, their violent deaths, and, the rise to power of Vespasianius (Haile 1958, 615). Octavia, as a real woman, had in actual historical fact been executed by Nero in 65 AD. In Anton Ulrich's novel, however, Octavia was rescued by the hero Tyridates, Prince of Armenia. In the novel the two are married in 70 AD, living a happy life thereafter as rulers of Media and Parthia (Haile 1958, 615). How this occurs will be seen in the new patterning of intentions and subsequent behaviors that establish the new normative patterns of effective ethical behavior, the significance of "being aware of how one is aware."

Typical of this tension between one's first response and how one alters it with a choice of an alternative perspective one sees in the hero Tyridates as he is faced with how to

---

3. H.G. Haile, "'Octavia: Römische Geschichte,': Anton Ulrich's Use of the Episode," *The Journal of English and Germanic Philology* 57, no. 4 (October, 1958): 616.
4. Hayden White, *The Fiction of Narrative, Essays on History, Literature and Theory, 1957–2007*, ed. Robert Doran (Baltimore: Johns Hopkins Press, 2010).

express himself toward a father-like figure who is in his filial emotional depths, when that person does something he has suddenly seen as unjust:

> Tyridates couldn't have encountered a greater discovery than what he now experienced. And even as he in each instance honored Volgeses as a father, he could not at this moment contain himself from writing to him of his awareness that he, Volgeses did not respect the Parthian marriage sufficiently.[5]

This being in the moment of two, then one mind is carried through every self-report in the book. Indeed, a narrative by a first person reflection is the structure of the book throughout. There is no objective, impersonal narrator, only the reflections of the differing personages that move the plot along. An example is of an unnamed minor figure in a period of Tyridates' life who narrates an early experience between himself, as observing narrator, Tyridates and Artabanus, Tyridates' cousin. The latter falls into a speechless love-sickness as Tyridates has evidently fallen in love with success with his own love interest. The narrator inserts himself into this "becoming aware of one's own awareness" as a key lever in this moment in the plot:

> Since both princes were fully open to each other since their childhood, Tyridates was quite perplexed as to why Artabanus would not speak with him, lying there for hours without any personal disclosure. But Artabanus thought he would rather die than reveal the cause of his pain. Yet his torture by this unfulfilled love was not long hidden from us. I was the first who discerned it, and decided to reveal my understanding. Finally, he recognized me, and admitted that I had puzzled out (erraten) the truth, upon which he painfully revealed that he wanted nothing better than to die, so that no disquiet awoke (erwekken) in his Tyridates, and that he could abide me no longer. (Ulrich 1993, 104)

One must note the focus on immediate moments of discovery and decision, so that the reader must engage his or her own such moments.

There is interestingly another level to the narrative, and that is something perhaps suggested by Leibniz who consulted with Anton Ulrich during the composition of *Octavia* (Haile 1958, 616 fn. 16; 626). This is the notion of the "inner becoming" that connects individual choice with what both Suarez and Leibniz understood as the "secondary" presence of God's will. There is, of course, no guarantee of this in the narrative, only the kind of speculation as the characters review their choices, that Pascal called his "wager." Indeed, the choice is not made so much as a "wager" of God's will, rather an attempt to comprehend surprising coincidences that eventuate from one's deliberative (as in the above examples) choices. Thus, while not as adamant as Hobbes in this separation of final and efficient cause in one's decisions in the uncertainty of the new secular world without signposts, Anton Ulrich, as Leibniz, nonetheless could speculate, but not be sure of why and how each decision fulfilled God's plan.

---

5. Anton Ulrich, *Die Römische Octavia,* Volume One, Part I, ed. Rolf Tarot and Maria Munding (Stuttgart: Anton Hiersemann, 1993), 113–14.

The formal cause dominant in this phase consists of the pattern of reflective choice upon differing self and other perspectives. A second aspect of that formal cause is a sense that all choices arises from interpersonal interaction that contribute to a final cause. In the understanding of Anton Ulrich, and Germany really, that of a secular fulfillment of God's "secondary" presence within the internal spirit of the person, his or her intrinsic *Energeia*.

C. Mathematics—The Further Development of Logical Calculus: Two Studies (1679)

Gottfried Wilhelm Leibniz (1646–1717)

> In 1679 Leibniz thought of the logical calculus as an application of the more general science of characters to the problems of formal logic. Such an application would, he was convinced, put logic on a more universal basis and serve to convince men of the value of applying symbols to material truth as well.
> Leroy E. Loemker, Gottfried Wilhelm Leibniz, *Philosophical Papers and Letters* (1969, 235).

Leibniz in these two studies creates a logical continuum that can be applied to a complete essay or argument that emerges from the material inquiry into a problem or issue. The material evidence with the insertion into a logical calculus generates a concept that enables a rational argument to be made from its material verification. With a logical calculus, it will have a sufficient extensive logical framework to guide explanations and, most importantly, further research into what can be seen as further evidence that is logically possible. Such a rational framework will enable a thinker, in any of the sciences, to formulate from the individual material fact a species—that individual which is examined, and couple it to conceptual arguments that give it a more universal basis that enables objective knowledge that can guide future findings. Each individual species will have a genus and differentias which compose a materially based argument that is continuous and extensive. With this logical calculus that relates subjects to predicables, inquiry that can be materially deeper, while possibly more limited in scope, can generate comprehensible and comprehensive argument for particular states-of-affairs. This, of course, will be the significance of material cause and well-tracked efficient causes in the third phase of the metaparadigm. In these second phasal two studies of logical calculus in 1679, Leibniz will address the formal cause of the value of the logical calculus, which is the sequence of steps required to fashion a coherent argument, using the logical calculus to keep track of the conceptual associations made.

> Two terms, one of which contains the other but which are not coincident are commonly called *genus* and *species*. Considered as concepts or component terms as I am here viewing them, these differ as part and whole, so that the concept of genus is part, the concept of species the whole, for it consists of genus and differentia. For example, the concept of gold and he concept of metal differ as a whole and part, for the concept of metal and something more is contained in the concept of gold; for example, the concept of the heaviest among the metals. Thus the concept of gold is greater than the concept of metal.

The schools speak otherwise, because they are considering not concepts but instances subsumed under universal concepts. Thus they say that metal is wider than gold, since it contains
more species than does gold. If we were to count the individuals made of gold on the one
hand, and those made of metal on the other, there would certainly be more of the latter than
the former, and hence the former would be contained in the latter as part of a whole. In fact,
by applying this observation and using fitting characters, we could demonstrate all the rules
of logic by another kind of calculus than the one developed here, merely by an inversion of
our own calculus. I prefer to consider universal concepts or ideas *and their composition* [my ital-
icization, M.E.B.], for these do not depend upon existence of individuals. So I say that gold
is greater than metal, because more constituents are required for the concept of gold than
to produce just a metal. Thus, our phrases here and the Scholastic phrases do not contradict
each other but must nevertheless be carefully distinguished. (Leibniz 1969, 237–38).

A logical calculus raises the material presence into a conceptual "species" interconnectable
to a conceptual genus that enables broader connections of thought. In problem solving,
my contention for all inquiry, genus connections are vital. A problem as a genus must be
broad enough to engage a range of minds in a discipline. We will see that in the fourth
phase of the metaparadigm, where complementary arguments are made from differing
perspectives that link the traditional to the new perspectives so as to find the normative answer. The addition of specific differentia to the material evidence of the species-
concept enables variables, such as context, temporality and other accidental factors to
be included in what one sees. These material identifiers, the predicables of a specific
argument, again can engage inquirers coming from differing perspectives to engage each
other. The use of a logical calculus helps one construct any such complex argument that
can be shared among inquirers.

Leibniz prefers the letters of alphabet to demarcate the parts of an argument, which proceeds sentential statement by sentential statement:

(1)   A *universal affirmative proposition* is here expressed in this way: *a* is *b*, or (all) man is
       animal. So we always understand the universal sign to be prefixed. We are not now
       discussing negative propositions or particular hypothetical propositions.

(2)   A proposition true in itself:
       ab is *a*, or (all) rational animal is animal.
       ab is *b*, or (all) rational animal is rational.
       Or omitting *b*, *a* is *a*, or (all) animal is animal.

(3)   *Conclusion true in itself:*
       If *a* is *b*, and *b* is *c*, then *a* is *c*. If (all) man is animal, and (all animal) is substance, then
       all man is substance.

(4)   From this follows:
       If *a* is *bd* and *b* is *c*. (All) man is rational animal; (all) animal is substance; hence (all)
       man is substance.
       This may be demonstrated as follows. If *a* is *bd*, by hypothesis, and *bd* is *b*, by no. 2,
       then *a* is *b* by no. 3. Further, if *a* is *b* (as we have proved), and *b* is *c*, by hypothesis,
       then *a* is *c*, by no. 3.

(5)   A *proposition is true* which arises through logical conclusions from given to given propositions which are true in themselves (1969, 240–41).

Using Leibniz's logical calculus, let *a* is *b*, or (all) formal cause is a sequential pattern of
possible inquiry. If *a* is *b*, then *a* is *c*. If (all) formal cause is a sequential pattern of possible inquiry, and all sequential pattern of possible inquiry furthers the systematic development of evidence. Hence all formal cause is the systematic development of evidence.
Then from this follows:

If *a* is *bd* and *b* is *c*. (All) formal cause is a sequential pattern of possible inquiry that
elicits the materially verifiable evidence of a concept. Hence all formal cause furthers the
systematic development of evidence.

This may be demonstrated as follows. If *a* is *bd* by hypothesis, and *bd* is *b* by
no. 3. Further, if *a* is *b* (as we have proved), and *b* is *c*, by hypothesis, then *a* is *c*, by no. 3.

Q.E.D.: *A proposition is true* which arises through logical conclusions from given to given
propositions which are true in themselves. My evidence for *a, b, c,* and *d* has been empirically verified by my citations from the texts of the thinker's themselves.

I could add to the above logical calculus proof the other characteristics I have
brought into my discussion insofar as how problems are both formulated and inquired
into in the metaparadigmatic phases. Besides the importance of the Aristotelian four
causes, I have emphasized the characteristic of "being aware of how oneself and others
are aware". This fulfills (*e*) the complex individual differences of all integrities (person,
place or thing) that participate in an assumed common totality. I have shown the
emphasis of thinkers in all three disciplines on individual choice, emerging from a view
of the autonomy of the person. This fulfills (*f*) the freedom of intelligent beings. I have
demonstrated in the thinkers the preference for definition, for the formation of meaningful concepts that are justified empirically, at least with some experiential evidence,
that seemingly interact with one another in a cause and effect manner. This will be (*g*)
mind rather than matter as the chief content of judgment. Finally, the inclination of
thinkers in all disciplines to make a sudden break with or radical re-definition of, or new
usage of the concepts of the inherited normative thought of an earlier metaparadigm.
This would be (*h*) quantum change in states-of-affairs. The systematic coherence of
(f) through (h) as a way of demonstrating the initial three phases of a metaparadigm is
my purpose in this text, aiding one in having evidentially a "configurational comprehension" of the metahistorical pattern. Again, the pattern is the Aristotelian formal
cause that guides a systematic inquiry.

Leibniz in this second phase, insofar as his development of differential calculus, developed in this calculus a mathematic method for determining an endless continuity of either
metaphysical or potentially empirical reality that could generate a focus for investigative
proofs. Differential calculus is of the spirit of the second phase of any metaparadigm—a
theory of systematic stages or developments that can become a basis for the subsequent
findings and definitional work that a logical calculus provides. Leroy Loemker writes of
this in the Introduction of his collection of Leibniz's essay and letters over the years of
the 1660s through 1716, shortly before his death:

... there is also the infinite involved in the possibility of endless analysis of the continuous finite. For every finite series involves infinite differentials of the first, second, and nth order and integrals similarly infinite. So too every perception involves an infinity of representative elements corresponding to the infinity of God's thoughts and their monadic expressions. Such infinity is a necessary consequence of the law of sufficient reason and the principle of plenitude or perfection, since no reason can be given for any finite limit. He writes in a note for a letter to Des Bosses:

There is a *syncategorematic* infinite or passive power having parts—namely the possibility of further progression in dividing, multiplying, subtracting, and adding. There is also a *hypercategorematic* infinite or potestative infinite, an active power having parts eminently, as it were, not formally or actually. This infinite is God himself. But there is no *categorematic* infinite, or one actually have infinite parts.

That is to say, Leibniz does not hold that there are real infinites and infinitesimals in existence, as Russell interprets him; they are in existence only as possibilities of analysis and synthesis. (Leibniz 1969, 30–31)

Loemker, later in this text, further clarifies this hypothetical infinite and infinitesimal as a guide for evidential research, essential in a systematic, interconnected set of hypotheses that may produce specific or suggestive empirical results:

It follows from this that even if someone refuses to admit infinite and infinitesimal lines in a rigorous metaphysical sense and as real things, he can still use them with confidence as ideal concepts which shorten his reasoning, similarly to what we call imaginary roots in the ordinary algebra, for example $\sqrt{-2}$. (1969, 543)

These "imaginaries" are for Leibniz necessary connectives that can guide one toward yet to be discovered, though never fully, evidential justifications. Leibniz connects this principle to his vision that all matter as it is realized in the forms of the world is in motion as in the Aristotelian vision of entelechy. Imaginaries can enable one to discern the presence of how the forms change, even as they are realized through the *Energeia* of the creative force of nature, be it its physical dynamics or the interventions of human thought. A systematic theory can guide inquiry into the *Energeia* in its steps and stages as it realizes the sought for form, the *Ergon*, yet the *Ergon* is always in potentially discernible transition: "*Forms* are for me nothing but activities of entelechies, and substantial forms are the primary entelechies" (1969, 511). The theory of a "spiral return" in human problemsolving perspectives, the metahistorical assumption of this text, can demonstrate how the *Ergon* of a particular stance in inquiry and explanation changes somewhat, based on what had been attempted in a previous metaparadigm.

Leibniz's use of the concept of "conation" justifies the Aristotelian view of the *Energeia* that realizes the form (the *Ergon*) even as it continues to change. Leibniz writes in this regard:

Conatus is to motion as a point to space, or as one to infinity, for it is the beginning and the end of motion. Hence *whatever moves*, no matter how feeble, and no matter how large may be the obstacle it meets, *will propagate its conatus in full against all obstructions* into infinity, and furthermore it will impress its conatus on all that follows. For though it cannot be denied that a moving body does not proceed in its motion even when it has been stopped, it at least strives to do so, and what is more, it strives, or what is the same thing, begins to move the obstructing bodies, however large, even though they may exceed it.

There can therefore be many contrary conatuses in the same body at the same time. For given the line *ab* and *c* moving from *a* to *b*, and *d*, on the other hand, moving from *b* to *a*, and colliding with *c*; then at the moment of collision *c* will strive against *b* even though it is thought to stop moving, because the end of motion is conation. (1969, 140).

One needs differential and integral calculus to determine the interactions of a conatus with another conatus. A logical calculus can indicate in its systematic development of certain possible state-of-affairs how to inquire for more information But, to know with evidence the definitive actuality of a certain concept of such a system, differential and integral calculus can indicate more exactly the scope of what is to be examined insofar as what, when, where and how. The logical calculus can then say "why." A logical calculus can examine, for example, the legislative interactions and outcomes of let us say, the writing of the American 1787 Constitution—considering the conatuses of mind and behavior of the 55 men in Philadelphia. Yet, a differential and an integral calculus can serve as even finer instruments for tracking the interactions. Statistical models of how larger and smaller groups function by certain efficiency standards are possible with mathematical calculi. New conceptual premises that are between or consequent to the current line of interconnected premises of a certain problem area—in this instance interpersonal cooperation—may have new insights when *a*, *b*, *c*...*n* are examined in their groupings and sequences of behaviors. The several hypotheses concerning the group interpersonal conatuses may be configurationally confirmed as actual in their details when, where and how, then a solidly supported "why." It is in the third phase of metaparadigms that these material and efficient causes are examined in the greatest depth and detail.[6]

---

6. A logical calculus, by itself, cannot justify the current working premise of the discipline of group dynamics training that a functional small group experience, where emphasis is placed upon interactions over a period of weeks, should be more than 12 persons and less than 20 persons. This size, ostensibly, is the manageable number of participants, if the interpersonal communications training is to become effective (see Morton A. Lieberman, Irvin D. Yalom and Matthew B. Miles, *Encounter Groups: First Facts*, New York: Basic Books, 1973), 118 [Table 3–8]. A differential and an integrative calculus could enable one to find an actual range of participants from its application to sufficient group samples. Not much has been written on this dimension of small group training, yet besides number, individual differences are also significant among membership. Paul Hare writes of these variables in *Handbook of Small Group Research* (New York: The Free Press, 1962), 387–96. Discussing the effects of interactive conatuses explicitly (without using the term conatus, rather "force field," the work and theory of Kurt Lewin is presented by Leland P. Bradford, Jack R. Gibb and Kenneth D. Benne in *T-Group Theory and Laboratory Method, Innovation in Re-education* (New York: John Wiley & Sons, 1964), 326–32.

Chapter Three

THE THIRD PHASE: MATERIAL INQUIRY INTO THE VERIFIABILITY OF SPECIFIC CONCEPTS, AND CONFLICT OVER THE IMPLICATIONS OF THE FINDINGS C.1690–C.1720

The conflicts between concepts in the third phase of the first Modern metaparadigm were between the traditional manner of understanding human existence and the new positions insofar as final cause, formal cause, material cause and efficient cause. The prominence of the Divine still resisted the separation of powers, as it were, between the human and the Divine. The stations of aristocracy, propertied commoner and commoner persisted. Every side of the debates sought to distinguish their thinking with helpmeets from tradition, yet also taking, when necessary, insights of the new modernism. Nonetheless, this time was characterized in all the disciplines as the "conflict between the ancients and the moderns." Writing on this of England in this period, Joseph M. Levine states in a book that takes up this controversy, which was in every nation of Europe in this time:

> In the following pages I have tried to characterize the high culture of Restoration England by concentrating on the broad and sometimes boisterous argument that broke out between the ancients and the moderns and that seemed for a time to have engaged and divided nearly everyone …
>
> Undoubtedly, there is something paradoxical in suggesting that it is through an argument that one may hope to characterize a culture … Unfortunately, the history of argument has not, I think, always had its proper place in the telling of intellectual, much less of cultural history.[1]

The leading minds who addressed this controversy in a way that created new modern insights in historical thought, literature and science knew the ancient, but challenged many of their precepts with their own new insights, drawn from empirical study. Below, an introduction to the new ideas:

> Slavery is so vile and miserable an Estate of Man, and so directly opposite to the generous Temper and Courage of our Nation; that 'tis hardly to be conceived, that an *Englishman*, much less a *Gentleman*, should plead for't. And truly, I should have taken Sr. Rt. Filmer's *Patriarcha*[2] as any other Treatise, which would perswade all Men, that they are Slaves … had not the

---

1. Joseph M. Levine, *Between the Ancients and the Moderns, Baroque Culture in Restoration England* (New Haven: Yale University Press, 1999), vii.
2. *Patriarcha* was written in 1637–38, but not published until 1680.

Gravity of the Title and Epistle, the Picture on the Front of the Book, and the Applause that followed it, required me to believe, that the Author and Publisher were both in earnest. I therefore took it into my hands with all the expectation, and read it through with all the attention due to a Treatise that made such a noise at its coming abroad. And I cannot but confess my self mightily surprised that in a Book, which was to provide Chains for all Mankind, I should find nothing but a Rope of Sand, useful perhaps to such whose Skill and Business is to raise a Dust, and would blind the People, the better to mislead them, but in truth is not of any force to draw those into Bondage, who have their Eyes open, and so much Sense about them as to consider, that Chains are but an ill wearing, how much Care soever hath been taken to file and polish them.

… His system lies in a little compass, 'tis no more but this,

That all Government is absolute Monarchy

And the Ground he builds on, is this,

That no Man is Born free.

In this last age a generation of men has sprung up among us, who would flatter princes with an Opinion, that they have a Divine Right to absolute Power … and by Divine Right, are Subjects to *Adam's* right Heir.

John Locke, Introduction to Two Treatises of Government (1690)[3]

In writing your Courants, we advise you carefully to avoid the Form and Method of Sermons, for that is vile and impious in such a Paper as yours. Here, perhaps you will say you do not set up for a *Preacher*, to which we Answer, that to print your Paper *Sermon-wise* is as bad as if you preach'd. And besides, for a private Man to Exhort and Admonish in such a method, is *boldly to invade the Province of others*, and comes little short of a *Corah-like* Usurpation. Nor is it suitable, as we conceive, to fill your Paper with Religious Exhortations of any kind; or to conclude your Letters with *the Words of the Psalmist*, or any other sacred write.

Benjamin Franklin (aged 16), a letter to the *New-England Courant*, under the pseudonym A, B, C &C., January 28, 1722/23[4]

About the middle of last winter, I went to see an opera at the theatre in the Haymarket, where I could not but take notice of two parties of very fine women, that had place themselves in the opposite side boxes, and seemed drawn up in a kind of battle array one against the another. [fn 1. Whoever recollects with what violence the spirit of party raged in the latter end of Queen Anne's reign, will not be surprised that it should infect the ladies, or show itself in the instances

---

3. John Locke, *Two Treatises of Government*, ed. Peter Laslett (Cambridge: Cambridge University Press, 1988), 141–42.
4. Benjamin Franklin, *Benjamin Franklin Writings* (New York: The Library of America, 1987), 41. Franklin wrote a host of letters with pseudonyms to this paper, where he worked as an apprenticed helper to his brother, James, who owned the paper. His brother did not know the identity of the letter writers. Letters such as these contributed to the arrest of James by the Chief Justice of the Massachusetts Supreme Court, Samuel Sewall. Benjamin defended his brother with a legal argument that Sewall enforced a law on James ex post facto for earlier issues of the *Courant* (1987, 47–48). Benjamin and his brother were at odds from the beginning of his apprenticeship, and Ben's letters had the intention to challenge his brother, besides the secular emphasis which I address above.

so pleasantly indicated in this paper; ed. George Washington Greene, 1854]. After a short survey of them, I found they were patched differently; the faces, on the one hand, being spotted on the right side of the forehead, and those upon the other on the left; I quickly perceived that they cast hostile glances upon one another; and that their patches were placed in these different situations, as party signals to distinguish friends from foes. In the middle boxes, between the two opposite bodies, were several ladies who patched indifferently on both sides of their faces, and seemed to sit there with no other interest than seeing the opera. Upon inquiry, I found that the body of Amazons on my right hand were Whigs, and those on my left Tories; and that those who had placed themselves in the middle boxes were of a neutral party, whose faces had not yet declared themselves. These last, however, as I afterwards found, diminished daily…

Joseph Addison, The Spectator, No. 81, Saturday, June 2, 1711[5]

"In fact," Leibniz said in his letter to Bernoulli, "For many years now the English have been so swollen with vanity, even the distinguished men among them that they have taken the opportunity of snatching German things and claiming them as their own." Bernoulli received Leibniz's letter and dashed one off in response, saying that his friend should consider proving the inferiority of the Brits by posing more challenge problems to them that could only be solved using calculus … The challenge was to determine the curve that should cut, at right angles, an infinity of curves expressible by the same equation. Unfortunately for Leibniz, this effort failed to reveal the inferiority of the English mathematicians because there was a problem with the way the challenge was written; it was interpreted to be asking for a specific example of such a curve rather than a general solution for finding such a curve—the much harder question Leibniz had intended.

On the Conflict in 1713 between Leibniz and Newton[6]

What all these extracts have in common is *conflict*. The conflict is either between an older final cause and a new one, such as Locke in his attack on Filmer's *Patriarcha*. The new, secular final cause is evident in Franklin's attack on the rhetoric of the *New England-Courant*. The final cause of governance is in conflict as evidenced in ordinary public life as Addison shows in his observations of public behavior, indicating the older monarchical sovereignty of the English Tories versus the emergent Whigs who represented commoners. The conflict had a new front increasingly in the new Modernist secularism that is between nations, where an emergent nationalism would increasingly affect what was deemed normative in the models of problem solving in the arts and sciences.

Conflict enters Western culture in a marked manner in the third phase of a metaparadigm, when competing systems of ideation in any discipline have formulated long-range projects of their approaches, and now enter more depth and detail in addressing specific problems with their methods. Material cause is one of the two guiding Aristotelian causes in this third phase. Material cause becomes prominent in the formulation of in-depth problems and approaches in physics, geology, botany, biology, the social

---

5. Joseph Addison, *The Spectator* in *The Complete Works of Joseph Addison*, Vol. V, ed. George Washington Greene (New York: G.P. Putnam, 1854), 225–26.
6. Jason Socrates Bardi, *The Calculus Wars, Newton, Leibniz, and The Greatest Mathematical Clash of All Time* (New York: Thunder's Mouth Press, 2006), 208–9.

and political contexts in their effects on everyday existence, human thought and new aesthetic ideas of the sublime, beauty and human form come into conflict—new ideas versus
new ideas in disciplines strive to command what should be normative inquiry and explanation. Efficient cause is also in the forefront of the third phase, building upon the focus
on individual agency and freedom of thought this first metaparadigm of Modernism has
stressed. Efficient cause of human agency is highly valued in this third phase. The role of
the individual in his or her affect upon the conduct of inquiry, but also as a change agent
or an agent who sustains tradition in an active manner against the revolutionary conatus
that would change the rights of everyday life insofar as the customs of the Estates. John
Locke, for example, will start a political party for the emergent republican institutions in
this phase, as will Henry St. John, Lord Bolingbroke in his ideational rejuvenation of the
monarchical vision. Both of these men will contribute not only to political philosophy,
but to the philosophy of history in ways that characterized the manner and significance
of individual action. Both men will come directly and continuously indirectly into the
severe conflicts of institutional change in England. The same story will be seen in France
between figures such as the Abbe St. Pierre and Montesquieu, the former an outspoken
liberal, inclined toward the rights of the ordinary person, and Montesquieu writing elo-
quently in defense of the executive authority of royalty. In Germany one sees the ideational conflict between the monarchist apologist, Anton Ulrich, and the strong voice for
at least a share of power from commoners like Christian Thomasius and Christian Wolff.
Such is the atmosphere of 1690 through 1720.

A. History/Philosophy of History

Anthony Ashley-Cooper, Third Earl of Shaftesbury (1671–1713)

The Third Earl of Shaftesbury was the grandson of the first Earl, Anthony Ashley
Cooper (1621–1683) who along with John Locke started the Whig Party in the English
Parliament in 1680. The Third Earl was a politically active Whig in the House of Lords
in the first decade of the eighteenth century. Yet, he was fair-minded, and can be likened
to those women who hadn't chosen a particular path who sat in the middle of the theater boxes, according to Joseph Addison (see above extract). This is important in that he
avoided pitched conflict and chose rather to engage in scholarship, his vision of historical
writing giving us the first historiographical observation in this initial metaparadigm of
Modernism. This in itself is of the third phase of the metaparadigm in that he addresses
a particular problem that had not been recognized in the writing of history, even while
apprehended by either proponents or opponents of what was written. This greater depth
can be said to have initiated the discipline of hermeneutics in the philosophy of history.
The conflict was to challenge all existing writers, most particularly his nuanced criticism of Thomas Hobbes' perspective.[7] Shaftesbury writes more generally of historical
perspectives that structure events:

---

7. Anthony Ashley-Cooper, The Third Earl of Shaftesbury, *Characteristicks of Men, Manners,
   Opinions, Times,* Vol. One (Indianapolis: Liberty Fund, 2001), 56–57.

And the historian or Relater of Things important to Mankind, must, whoever he may be, approve himself many ways to us; both in respect of his Judgment, Candor, and Disinterestedness, e'er we are bound to take any thing on his Authority. And as for "*critical Truth*", or the Judgment or Determination of what Commentators, Translators, Paraphrasts, Grammarians, and others have, on this occasion, deliver'd to us in the midst of such variety of Style, such different Readings, such Interpolations, and Corruptions in the Originals; such Mistakes of Copists, Transcribers, Editors, and a hundred such Accidents, to which ancient Books are subject: it becomes, upon the whole, *a Matter of nice Speculation*, considering, withal, that the Reader, tho an able Linguist, must be supported by so many other helps from Chronology, natural Philosophy, Geography, and other Sciences.

And thus many previous *Truths* are to be examin'd, and understood, in order to judge rightly of *historical Truth*, and of the past Actions and Circumstances of Mankind, as deliver'd to us by antient Authors of different Nations, Ages, Times, and different in their Characters and Interests. (Shaftesbury 2001, 91–92)

The Third Earl of Shaftesbury is concerned, as well, with how the agents of historically relevant action, and otherwise, engage others. His views on personal character insofar as both public and private interrelationships, goes further in this attention to individuality that is one of the hallmarks of Modernism. His approach emphasizes a balance of competing emotions that enables the person to be objective, humane and understood by others. His listing of the several competing emotions is an augmentation of Hobbes, giving us more depth and detail on human thought and its responses to and affects upon emotional life:

Remember, therefore, in manner and degree, the same involution, shadow, curtain, the same soft irony; and strive to find the character in this kind according to proportion both in respect of self and times. Seek to find such a tenour as this, such a key, tone, voice, consistent with true gravity and simplicity, though accompanied with humour and a kind of raillery, agreeable with a divine pleasantry.—This is a harmony indeed! What can be sweeter, gentler, milder, more sociable, or more humane? Away, then, with that other sociableness; that inwardness, intimacy, openness. How false, how unfounded, how harsh in reality, and unfitted for what it is designed; how unfitted for their good whom it is meant to serve, and for thine, in respect, of thy own character, conviction, improvement! Indeed the very reverse of all.

But truth! Truth!—Remember that truth is best preserved when those thou conversest with are made to think most truly of thee; and this will least be when thou speakest, most truly or most simply in this way, or wouldst correct, rebuke, and teach with the same simplicity. Seek, then, the true simplicity: for this thou usest with them is not so. As for gravity, used in their concerns, as hoping or expecting better of them, this in good earnest ridiculous; and not only that, but in another respect tyrannical and barbarous.

Firm, steady, even, upright, between these contrary blasts, efforts of humour, temper, sallies of disposition, the gay, light-wings zephyrs, and the ruffling Boreas or heavy Notus—Colossus-like, fixed, poised with equal footing and foundation on each side—a promontory parting two seas ... Firm, steady, etc.—Equal between these two extremes of different brows. Both mixed in a manner; convertible, communicable by an easy change from one into another; not starting, not shrinking from one another; not constituting two different souls, two different men, differently known, differently accessible, differently to be tested, spoken with ...

> Earnest: but not in earnest. Jest: but not really jest … Never to leave till this balance be brought tight, or pretty near to evenness …[8]

The Third Earl of Shaftesbury narrates his advice as if he is in the midst of a conversation, perhaps with more than one person. He does not cogently formulate a short principle, and then cease or continue to the next issue. We will see this style in our purview of literature in this third phase. The material and efficient causes highlighted are the actual context of a conversation and the shifting requirement to integral two tones into a rhythmic engagement of one or more others.

We will see this more conversational philosophy of history in John Locke (1632–1704), a mind so prominent even today in political, but also in our existing laws that treat what is considered the normal individual, as opposed to the irrational one. This latter significance is in his definition of the inability to know right from wrong in a moment of passion, his thoughts on madness, one of the first bases of the Modernist insanity plea.[9] This concern with the nature of individual thought will be considered as integral to a philosophy of history, for as the Third Earl of Shaftesbury, who was tutored by Locke, Locke is concerned with how cultural-historical change occurs, and can occur. This latter concern, how individuals can be educated into new concepts of normative political-social reality, opens up a new discipline of pedagogy, planned experiential education:

> Let us then suppose the mind to be, as we say, white paper, void of all characters, without any ideas:—How comes it to be furnished? Whence comes it by that vast store which the busy and boundless fancy of man has painted on it with an almost endless variety? Whence has it all the MATERIALS of reason and knowledge? To this I answer, in one word, from EXPERIENCE. In that all our knowledge is founded; and from that it ultimately derives itself. Our observation employed either, about external sensible objects, or about the internal operations of our minds perceived and reflected on by ourselves, is that which supplies our understandings with all the MATERIALS of thinking. These two are the fountains of knowledge, from whence all the ideas we have, or can naturally have, do spring. (Locke 1962, 25–26)

This experiential focus suits the dominant efficient cause of any third phase. This is a time of trial and error. The time of active in depth inquiry into the ins and outs of the systematic theory developed in the second phase of the metaparadigm. The experiential experiments are into how humans learn, challenging the older, traditional view of "innate" ideas, the view of a predestined, divine final cause intrinsic to each individual. Yet, Locke will observe himself differences that separate each child from another; however, he will not credit that with predestination. Rather, all children each have interests, natural tendencies. A sound experiential education is to allow them to follow their own personal

---

8. Anthony Ashley-Cooper, The Third Earl of Shaftesbury, "Character," in *The Life, Unpublished Letters and Philosophical Regimen of Antony Earl of Shaftesbury*, ed. Benjamin Rand (London: Swan Sonnenschein, 1900), 193–94.

9. John Locke, *Locke's Essay Concerning Human Understanding*, Books II and IV, ed. Mary Whiton Calkins (LaSalle, Ill.: Open Court, 1962), 87–88.

interests, and as their parents or teachers, facilitate those choices with knowledge and encouragement.

> For Locke, educating children, then, entails instructing their minds and molding their natural tendencies. Education develops the understanding, which men "universally pay a ready submission" to, whether it is "well or ill informed" (CU: § 1). Because children are born without a natural knowledge of virtue, early education greatly shapes their development, where even "little and almost insensible impressions on [their] tender infancies have very important and lasting consequences" (TCE: § 1). Thus, Locke's method of education is meant to be observed by parents even from the time their child is in the cradle, long before the teaching that comes from books. Locke warns at the end of Some Thoughts Concerning Education that he can only provide general views on the proper education of a gentleman; the "various tempers, different inclinations, and particular defaults that are to be found in children" are so diverse that "it would require a volume" to prescribe correct remedies for all (TCE: § 216). Locke encourages parents to watch their children, for through observation, parents can understand their child's distinctive inclinations. Specifically, they should pay particular attention to their child "in those seasons of perfect freedom" and "mark how [the child] spends his time" (TCE: § 125). Once armed with such information, parents can better know how to motivate their children towards the right and can craft their methods of education accordingly. As Nathan Tarcov notes, "One studies a child's nature not merely to adjust to it but to see 'how it may be improved'."[10]

Locke, as is the wont in the third phase of a metaparadigm, creates in his narrative style what I will call going forwards a "haptic" contact with the reader.[11] Not only will the narrative be more conversational (which literature will bring new emphases to), but the reader is encouraged to do what the narrator himself is doing. As he speaks of how immediate intuition of what is before one supersedes a conceptual demonstration, he writes:

> ... no more than it can be a doubt to the eye (that can distinctly see white and black), whether this ink and this paper be all of a colour. (Locke 1962, 277)

And, he even more in a "haptic" narrative style engages the reader in a bodily gesture:

> Self is that conscious thinking thing ... which is conscious of pleasure and pain, capable of happiness or misery, and so is concerned for itself, as far as that consciousness extends. Thus

---

10. Jamie Gianoutsos, "Locke and Rousseau: Early Childhood Education," *The Pulse, the Undergraduate Journal of Baylor University* 4, no. 1 (2006): 2.
11. The word came into usage in the final decades of the seventeenth century, the third phase of the first Modernist metaparadigm. There was the realization by philosophers like Locke, and also artists that images which were frontal illusions that seemed to reach out to the reader or viewer were comprehended by a sense of contact. The Oxford English dictionary defines "haptic" as "tactile and kinaesthetic sensation; touch, esp. as a means of nonverbal communication." Origin: A borrowing from post-classical Latin *haptice* (1685) as quoted from a text in 1734 by J. Kirby.

every one finds, that whilst comprehended under that consciousness, the little finger is as much part of himself as what is most so. Upon separation of this little finger, should this consciousness go along with the little finger, and leave the rest of the body, it is evident that the little finger would be the person, the same person ... (Locke 1962, 255)

Locke compels the reader to examine his or her own verbal narrative style, asserting that one's individuality lies in one's choice of grammatical expression. This is *the* onset of grammatical stylistics, taken up by Leibniz in his reading of this Lockean text—in this same third phase. The Modernist metaparadigms have not gone beyond this depth of individual insight into individual expression. What Locke says below is akin to the discovery of not only fingerprints that prove absolute individuality in the late nineteenth century (the same third phasal stress on in-depth inquiry into individuality), but the discovery of complex DNA evidence that attests to extreme individuality in the third phase of the Modernist metaparadigm we are now in. Leibniz writes, quoting Locke in part:

A man must reflect on his own thoughts, and observe "the postures of his mind in discoursing"; for particles are all marks of the action of the mind.[12]

Locke's is aware of how the mind is a "blank slate" can be written upon by a focused pedagogy. His advice to parents in observing the individuality in interests and actions of their children, and, his focus upon the radical new evidence for individuality in one's "postures of his mind in discoursing," serve Locke in his desire for historical change in the institutions of the state toward democracy. He writes of this natural equality of all, and its implications for governance, in his *Two Treatises of Government*:

To understand Political Power right, and derive it from its Original, we must consider what State all Men are naturally in, and that is, a *State of Perfect Freedom* to order their Actions and dispose of their Possessions, and Persons as they think fit, within the bounds of the Laws of Nature, without asking leave, or depending upon the Will of any other Man.

A *State* also *of Equality*, wherein all the Power and Jurisdiction is reciprocal, no one having more than another: there being nothing more evident, than that Creatures of the same species and rank promiscuously born to all the same advantages of Nature, and the use of the same faculties, should also be equal one amongst another without Subordination or Subjection ... And Reason, which is that Law, teaches all Mankind, who will but consult it, that being all equal and independent, no one ought to harm another in his Life, Health, Liberty, or Possessions. (Locke 1988, 269, 271)

It is well-known, of course, that Locke felt that from "experience" one who had property learned a prudence and ability to sustain and increase personal wealth that made

---

12. Gottfried Wilhelm Leibniz, *New Essays on Human Understanding*, ed. Peter Remnant and Jonathan Bennett (Cambridge: Cambridge University Press, 1996), 330. Leibniz quotes Locke's *Essay Concerning Human Understanding*, Part III, Par. 4.

him more responsible not only as a self-governor, but a governor for others to acquire and maintain "life, liberty, and possessions." Locke was one of the major contributors to republican principles of governance, as well as in his efforts to facilitate that form of governance in the England of his time. Yet, insofar as the focus upon individualism, he was consistent in his contributions to a societal change based upon democratic principles where only the constructive value for self-development in the possession of property separated him from a thinker like Rousseau, whom we turn to in the second Modernist metaparadigm, when discussing philosophy of history.

B. Literature

Daniel Defoe (1660–1731)

> Honest John [Bunyan author of *Pilgrim's Progress* (1678)— M.E.B.] was the first I know of who mix'd Narration and Dialogue, a Method of Writing very engaging to the Reader, who in the most interesting Parts finds himself as it were brought into the Company & present at the Discourse. Defoe in his Cruso, his Moll Flanders, Religious Courtship, Family Instructor, & other Pieces, has imitated it with Success. And Richardson has done the same in his Pamela, &c.—. (Benjamin Franklin, *Autobiography*, 1987, 1326)

John Bunyan ushered in the second phase of Modernist literature with the life story of Christian, who in an everyday environment confronts the seven deadly sins. The sins are presented as persons with that sinful caste of mind who speak to him in everyday language. This enables the reader, as Franklin attests, to "be brought into the Company & present at the Discourse" is improved upon by the depth and detail of Daniel Defoe's third phasal novels. Defoe's novels, *Robinson Crusoe* (1719) and *Moll Flanders* (1722) deepen with detail the fluctuations of the narrator's mind as they reflect upon their immediate circumstances. Individuality deepens beyond the rights and wrongs of Christianity. Rather, the secular world as a sociological reality now enters narrative complexity, Defoe in Moll Flanders showing how one's very dialogical norms change within her attempts to raise her primary partner to the level of societal behavior, beyond that of personal passion:

> All this while he lay at my Mother to find out, if possible, what was the meaning of that dreadful Expression of mine, as he call'd it, which I mention'd before; namely, *That I was not his lawful Wife, nor my Children his legal Children*: my Mother put him off, told him she could bring me to no Explanations, but found thee was something that disturb'd me very much, and she hop'd she should get it out of me in time, and in the mean time recommended to him earnestly to use me more tenderly, and win me with his usual good Carriage; told him of his terrifying and affrighting me with his Threats of sending me to a Mad-house and the like, and advis'd him not to make a Woman Desperate on any account whatever.

> He promis'd her to soften his Behaviour, and bid her assure me that he lov'd me as well as ever, and that he had no such design as that of sending me to a Mad-house, whatever he might say in his Passion; also he desir'd my Mother to use the same Perswasions to me too, and we might live together as we us'd to do.

I found the Effects of this Treaty presently; my Husband's Conduct was immediately alter'd, and he was quite another Man to me ...[13]

Moll's normative behavior strives for a "social contract." Indeed, we will see in Defoe's other works the emphasis upon the new middle-class station that has emerging rights. Public and private behavior becomes standardized for each station. This reflects the Third Earl of Shaftesbury's caution against too much openness, too much subjective honesty. This is not a loss of genuine individuality, rather an added choice in behavior. The use of the nouns of "Carriage" and "Treaty" mark for the reader a similar caution in conversation as with Shaftesbury. In fact, Defoe almost plagiarizes these sentiments found in Shaftesbury, but co-opting them for the middle-class. Defoe in *Robinson Crusoe* has Crusoe recognize the new rights and responsibilities of middle-class "station":

He bid me observe it, and I should always find, that the calamities of life were shared among the upper and lower part of mankind; but that the middle station had the fewest disasters, and was not exposed to so many vicissitudes as the higher or lower part of mankind; nay, they were not subjected to so many distempers and uneasinesses either of body or mind, as those were who by vicious living, luxury and extravagances, on the one hand, or by hard labor, want of necessaries, and mean or insufficient diet on the other hand, bring distempers upon themselves by the natural consequences of their way of living: that the middle station of life was calculated for all kind of virtues and all kinds of enjoyments, that peace and plenty were the handmaids of a middle fortune; that temperance, moderation, quietness, health, society, all agreeable diversions, and all desirable pleasures, were the blessings attending the middle station of life ...[14]

One of the great pleasures in reading *Robinson Crusoe* is the depth of detail Defoe provides in describing the everyday life of the marooned Crusoe. One can find this on every page, but at random I offer the following:

Be pleased to take a sketch of my figure as follows:

I had a high shapeless cap, made of goatskin, the skirts coming down to about the middle of my thighs; and a pair of open-kneed breeches of the same; the breeches were made of the skin of an he-goat, whose hair hungdown such a length on either side that, like pantaloons, it reached to the middle of my legs; stockings and shoes had I none, but had made me a pair

---

13. Daniel Defoe, *The Fortunes and Misfortunes of the Famous Moll Flanders, &c*, Who was Born in Newgate, and during a Life of continu'd Variety for Three-score Years, besides her Childhood, was Twelve Year a Whore, five times a Wife (whereof once to her own Brother) Twelve Year a Thief, Eight Year a Transported Felon in Virginia, at last grew Rich, liv'd Honest, and died a Penitent. Written from her own Memorandums. Volume One (Oxford: Basil Blackwell, 1927), 102–3.
14. Daniel Defoe, *Robinson Crusoe* (New York: Bantam Books, 1981), 2–3.

of somethings, I scarce know what to call them, like buckskins, to flap over my legs, and lace on either side like spatterdashes; but of a most barbarous shape, as indeed wee all the rest of my clothes.

I had on a broad belt of goatskin dried, which I drew together with two thongs of the same, instead of buckles; and in a kind of a frog on either side of this, instead of a sword and a dagger, hung a little saw and a hatchet, one on one side, one on the other. I had another belt not so broad and fastened in the same manner, which hung over my shoulder; and at the end of it, under my left arm, hung two pouches, both made of goatskin too; in the one of which hung my powder, in the other my shot. At my back I carried my basket, on my shoulder my gun, and over my head a great clumsy ugly goatskin umbrella, but which, after all, was the most necessary thing I had about me, next to my gun. As for my face, the color of it was really not so Mulatto like as one might expect from a man not at all careful of it and living within nineteen degrees of the equinox. My beard I had once suffered to grow till it was about a quarter of a yard long; but as I had both scissors and a razor sufficient, I had cut it pretty short, except what grew on my upper lip, which I had trimmed into a large pair of Mahometan whiskers, such as I had seen worn by some Turks who I saw at Sallee; for the Moors did not wear such, though Turks did; of these mustachios or whiskers, I will not say they were long enough to hang my hat upon them, but they were of a length an d shape monstrous enough and such as in England would have passed for frightful. (Defoe 1981, 134–35)

C. Mathematics—A Study in the Logical Calculus, Early 1690s

Gottfried Wilhelm Leibniz (1646–1717)

This study is one that adds more depth and detail to a particular aspect of the logical calculus of propositional truths that will be verbal, but with logical calculus analysis can be denoted in its phrases and clauses as its argument is made. The mathematical problematics are what is identical, i.e., coincident in two propositions; what is contained in one, that is subsumed by it (which we now call a thesis and sub-thesis), and what is a combination of two propositions where part of that combined statement contains identical or coincident parts, but are separate and not wholly identical in themselves.

A, B, or C, and D, and R, X, S serve to symbolize these several propositional truths, as in his 1679 two studies. However, Leibniz adds a visual dimension that enables in its figuration what I have called a "haptic" relationship for the reader. The imagistic presence of a visual design creates spatial access that offers itself in a way that a finger can trace the parts that are contained, coincident, or in part, identical and in part, not.

In the visualization below, Leibniz demonstrates how A and C are not coincident, i.e., equal, but they are both contained in RX. B is coincident with A, and it too is contained within RX. B is not coincident with C. RS and SX are not coincident.

The containing proposition is RX, which has non-coincident parts of A, C, but the coincident parts AB (Leibniz 1969, 373).

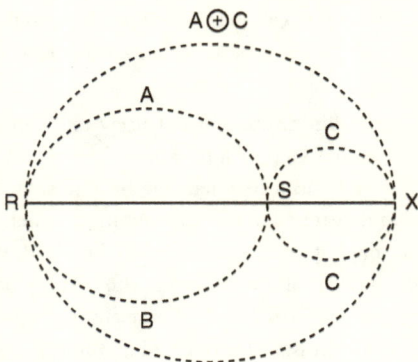

This can be written as the following set of propositions. RX is a thesis that states "All individuals are free to choose how they live." SX is a sub-thesis that states "Some individuals choose to break the laws of their society." A is a sub-thesis that states "Some individuals choose to follow the rules with enthusiasm." B is a sub-thesis that states "Some individuals choose to follow the rules less enthusiastically." Another sub-thesis is AB "Most individuals choose to follow the rules of their society." From this visualization one can begin to ponder the issue of freedom to choose how one lives within society. Various positions can then be taken. Recall that Hobbes felt that given the fundamental freedom of each person, they were responsible to themselves (and this principle) to escape from prison. The visual guide can keep this set of propositions in mind in a manner that a painting can be a constant guide to certain normative understandings of a problem, the terms of the problem and a prod to solve the problem.

Chapter Four

THE FOURTH PHASE: INTEGRATING THE NEW FOUR CAUSAL UNDERSTANDINGS WITH THE TRADITIONAL C.1720–C.1750

I think enough has been already said, to establish "the first and true principles" of monarchical and indeed "of every other kind of government;" though Mr. Locke condescended to examine those of Filmer, more out of regard to the prejudices of the time, than to the importance of the work. Upon such foundations we must conclude, that since men are directed by nature to form societies because by their nature they cannot subsist without them, nor in a state of individuality. And since they were directed in like manner to establish governments, because societies cannot be maintained without them, nor subsist in a state of anarchy; the ultimate end of all governments is the good of the people, for whose sake the were made, and without whose consent they could not have been made. In forming societies, and submitting to government, men gave up part of that liberty to which they are all born, and all alike. But why? Is government incompatible with a full enjoyment of liberty? By no means.

Lord Bolingbroke, On The Idea of The Patriot King (1738)[1]

The more tranquil the state of the body the more capable it is of portraying the true character of the soul. In all positions too removed from this tranquility, the soul is not in its most essential condition, but in one that is agitated and forced. A soul is more apparent and distinctive when seen in violent passion, but it is great and noble when seen in a state of unity and calm. The portrayal of the suffering alone in Laocoon would have been *parenthysros;* therefore the artist, in order to unite the distinctive and the noble qualities of soul, showed him in an action that was closes to a state of tranquility for one in such pain …

The common taste of artists of today, especially the younger ones, is in complete opposition to this … The arts themselves have their infancy as do human beings, and they begin as do youthful artists with a preference for amazement and bombast. Such was the tragic muse of Aeschylus; his hyperbole makes his Agamemnon in part far more obscure than anything Heraclitus wrote. Perhaps the first Greek painters painted in the same manner that their first good tragedian wrote.

---

1. Henry St. John, Lord Bolingbroke, *On The Idea of a Patriot King* in *Letters on the Study and Use of History* (London: Ward, Lock & Co., N.D.), 204.

Rashness and volatility lead the way in all human actions; steadiness and composure follow last.

Johann Joachim Winckelmann, Reflections on the Imitation of
Greek Works in Painting and Sculpture (1755)[2]

Sarah: Marwood will not escape her fate; but neither you nor even my father shall be one who condemns her. I die and forgive the hand which fulfills God's desire to have me home.—Oh, my father, what dark pain overcame her?—I still love you, Mellefont, and if loving you is a crime, how guilty I will appear in the other world!

Gotthold Ephraim Lessing, Miss Sara Sampson, A Middle-Class
Tragedy in Five Acts (1755)[3]

If to draw characters *justly*, and to support them *equally*:

If to raise a distress from *natural* causes, and to excite compassion from *proper* motives:

If to teach the man of *fortune* how to use it; the man of *passion* how to subdue it; and the man of *intrigue*, how gracefully, and with honour to himself, to *reclaim*

... If these, embellished with a great variety of entertaining incidents be laudable or worthy recommendations of any work, the editor of the following letters, which have their foundation in *truth* and *nature*, ventures to assert, that all these desirable ends are obtained in these sheets ...

Samuel Richardson, Preface, *Pamela; or, Virtue Rewarded* (1740).[4]

If it is true that the character of the spirit and the passions of the heart are extremely different in the various climates, *laws* should be relative to the differences in these passions and to the differences in these characters.

Montesquieu, On the laws in their relation to nature and climate in
The Spirit of the Laws (1748)[5]

The fourth phase of any metaparadigm gives resonance to Johann Joachim Winckelmann's dictum that while new ideation is challenging, indeed, in his mind too bold and brazen, given tradition, that finally thinkers realize how to realize their ideation with calm and composure so that it might realize its aims. The intention of this fourth cultural phase of the metaparadigm is to integrate the old ideation with the new so that an in-common society with cooperative inquiry can be achieved to a greater degree, so that long-range

2. Johann Joachim Winckelmann, *Reflections on the Imitation of Greek Works in Painting and Sculpture*, trans. Elfriede Heyer and Roger C. Norton (La Salle, Ill.: Open Court, 1987), 35, 37.
3. Gotthold Ephraim Lessing, *Miss Sara Sampson, Ein bürgliches Trauerspiel in fünf Aufzügen* in *Lessings Werke in Fünf Bänden* (Berlin and Weimar: Aufbau Verlag, 1964), 1: 92.
4. Samuel Richardson, *Pamela; or, Virtue Rewarded* (New York: New American Library, 1980), 21–22.
5. Montesquieu, *The Spirit of the Laws*, trans. and ed. Anne M. Cohler, Basia Carolyn Miller and Harold Samuel Stone (Cambridge: Cambridge University Press, 1989), 231.

solutions to the problems addressed in the disciplines of the arts and sciences may enable the society to truly foster progress.

Thus, we see in the above epigraph of Lord Bolingbroke an attestation to the principle that there is no individual possible without society. And, that laws that demand some sacrifice by the individual in his or her autonomy (as demanded by all thinkers in the previous three phases of the First Modernist metaparadigm) are necessary would an in-common government with established authorities guide the populace toward positive individual and collective expression and behavior.

The idea of an in-common unity pervaded the composition of novels, drama, painting and the development of sciences that began to insist upon one objective truth. Voltaire ponders as early as the 1730s the significance of what he calls "the three unities" of time, place and manner in drama. With these unities, the heroes and heroines, as opposed to the villains, modeled the composure sought by Winckelmann and Richardson. The contrast to this had been, and will be in the initial three phases of a metaparadigm, the emboldened challenge to the status-quo that is anathema to the fourth phase. We will see artists living and writing in this time, most markedly Gotthold Ephraim Lessing, shift their thought to what became the Storm and Stress period in Germany, with discontinuities in character and plot being prized. Without the integration of time, place and manner, throughout all aspects of the presentation, Voltaire will assert that the audience cannot comprehend the role of society in generating social laws of justice and mutual cooperation, thus, cannot appreciate how villainy does more than harm an individual, it either retards or disrupts a culture. He writes in the renowned essay *On Taste* in Diderot's 1755 *Encyclopedia*:

> When there is little sociability, the mind shrinks and grows dull because there is nothing to educate its taste. When some of the fine arts are absent, the others rarely manage to exist, because all the arts are interdependent, and sustain each other.[6]

Voltaire's stress on the "interdependence" of the arts occurs in a fourth phasal reality, as this is a time when there is more conscious focus upon a univocal reality achieved through the sharing of principles, even in separate disciplines. Voltaire also engages in opera, a model for this interdependence of the arts. In a *spiral* return to a fourth phase in the nineteenth century, c.1820–1867, opera thrived, where the cooperation of set design, staging and other minor arts combine. The return is spiral in that it builds upon what was achieved in the previous fourth phase, a phase where Voltaire's staging of drama, along with Gottsched as well as at that time Lessing in Germany, and Lillo in England, presented the *la belle natur* of the three unities of time, place and manner. And the *spiral* return of the fourth phase in the twentieth century was spurred by the Bauhaus of the 1920s and 1930s, where this interdependence of the arts was stressed in drama and dance, painting and stage setting. Among fascist playwrights, one can see this unity of

---

6. Alexander Gerard, *An Essay on Taste* (New York: Garland, 1970), 216–17 [Gerard's translation departs from other translations of the Encyclopedia]. See Diderot, *Encyclopedia*, 340.

time, place and manner, in Pirandello's (1921) *Six Characters in Search of an Author*. Both the Bauhaus and the fascists added new twists to that unity of time, place and manner, even using the older form of opera as a foundation for new expressive aesthetics to engage the reflective participation of an audience.

Voltaire champions the significance of "reflection" in this essay, which is the "being aware of how one is aware." He argues that such self-understanding, and personal change of character, cannot come without the artist's presentation of the unity of time, place and manner. Such a unity manifests a standard of a comportment among the characters to be emulated that cannot be properly understood without the unity of action and reaction throughout the drama:

> A young man who is sensitive but untutored cannot at first distinguish the parts in a large chorus; in a painting, his eyes do not at first distinguish the shadings, the chiaroscuro, the perspective, the harmony of its colors, and the correctness of the draughtsmanship; yet little by little his ears learn to hear and his eyes to see. The first time he sees a beautiful tragedy he will be moved, but he will be unable to discern either the effect of the unity, or the subtle art by which all unjustified entrances and exits are avoided or the even greater art by which unity of interest is created, or any of the other difficulties mastered by the author. Practice and reflection alone will make it possible for him to experience immediate pleasures from elements that formerly he could not distinguish at all. Good taste develops gradually in a nation that has hitherto lacked it because, little by little, men come under the influence of good artists. (Voltaire in Gerard 1970, 211–12; in Diderot, *Encyclopedia*, 338)

Benjamin Franklin's experience in the early 1720s is called to mind as one sees the significance of interpersonal integration, of mutual understanding, or finding peaceful solutions, of achieving composure in one's comportment. Franklin writes of this in his *Autobiography*, recalling these years in which in which Alexander Pope's (1688–1744) words on finding interpersonal agreement meant so much to him:

> Men should be taught as if you taught them not,
> And things unknown propos'd as thing forgot,—
> To speak th'o sure, with seeming Diffidence.
> Immodest Words admit of *no* Defence;
> *For* Want of Modesty is Want of Sense. (Franklin 1987,
> 1322)

Franklin adds that in these years he began to practice such diffidence, when speaking with others in a change of opinions or views, he used words that would defuse situations:

> … never using when I advance any thing that may possibly be disputed, the Words, *Certainly, undoubtedly,* or any others that give the Air of Positiveness to an Opinion; but rather say, *I conceive,* or *I apprehend* a Thing to be so or so, *It appears to me,* or *I should think it so or so for such & such a reason,* or *I imagine* it to be so, or *it is so if I am not mistaken.* (Franklin 1987, 1321–22)

It is in this fourth phase that the general principles which govern inquiry and expla-nation are articulated that will endure for decades, offering a standard language for the univocal idea of objectivity. In the sciences, supported by a view, the human being is a determined creature, as all animate beings, by their environment. Montesquieu will contribute to biology, meteorology and other auxiliary sciences in what in the next century will be the *spiral* return of such a fourth phase (c.1820–1865), where in-common principles stress the determinism of natural environs and organic structure in shaping all animate life. Charles Darwin (1809–1882) will be more informed by Montesquieu's research and explanation of the material cause of environment than his own grandfather, Erasmus Darwin. Erasmus Darwin (1731–1802) worked in the initial three phases of the Second Modernist metaparadigm, and his writings reflect the "intentionality," the choice, of natural agents in furthering evolution. Charles Darwin's further development of the material and formal causes of environment in species development, as well as his philosophy of final cause, further distanced from the Divine, even as a "secondary" or "proximate" cause, was still shared by most of his contemporaries.

Thus, as we take up our representatives of philosophy of history and historical writing, literature and science in this final phase of the first metaparadigm, the collective expres-sion of persons, places and things as a totality formed in common by dint of shared characteristics will be evident (rather than a focus upon the individual in greater depth). The determinism in the laws of function in intelligent beings and in things without mind or will (rather than the previous stress on intentionality, in the initial three phases) will be the converse focus of this final phase. A concept of duration of states-of-affairs (rather than the previous focus on quantum change), due to the subject now being groups, na-tions and species, where time is measured in longer segments; and, matter rather than mind as the chief content of judgment. In this fourth characteristic of the fourth phase of all metaparadigms, what is studied in the physical constitution of species as affected by the physical laws of nature. While reflection, "being aware of how one is aware" is present, the new awareness of one's awareness is more that of a pre-sociology, of how one is conditioned by society. And, as we see the impact of the new, fourth phasal sciences of nature, in our not too distant past, we learned how our natural state of being is deter-mined in ways never before recognized.

A. Philosophy of History/Societal Development of Political Institutions

Charles-Louis de Secondat, Baron de La Brède et de Montesquieu (1689–1755)

The mores of the prince contribute as much to liberty as the laws; like the laws, the prince can make beasts of men and men of beasts. If he loves the free souls, he will have subjects; if he loves common souls, he will have slaves.

Montesquieu 1989, 204

Montesquieu considered education as necessary for each "station" of society would its members find sufficient freedom for their milieu in an intelligent and cooperative spirit. Unlike Locke, he did not view the human mind as a *tabula rasa,* although he did not take up such cognitive issues in the depth of a Locke. Rather, he favored the tradition of his own aristocratic class, a class of people who when they fulfilled their traditional mission, cared for those of the common classes. Thus, Montesquieu wished an educated populace, especially in the laws, whose existence gave them protection of their liberties, and were a tool with which to lever their relation to aristocracy. Montesquieu took up Bolingbroke's notion of a "balance of power" between the role of aristocracy in governance and the role of the emergent "station" of commoners, in this instance, republicans. Rousseau's significant writings on behalf of those without property were a few years in the future (1756, 1762). Montesquieu's *The Spirit of the Laws* (1748) was in the spirit of the integrative intentions of this fourth phase of knowledge, and thus tradition plays a significant role in his arguments. Montesquieu defended the right of monarchs to be the executive authorities in a government, as well as other aristocrats who would be the Ministers of State. He writes in *The Spirit of the Laws*:

In monarchies the principal education is not in public institutions where children are instructed; in a way, education begins when one enters the world. The world is the school of what is called *honor,* the universal master that should everywhere guide us.

Here, one sees and always hears three things: that *a certain nobility must be put in the virtues, a certain frankness in the mores, and a certain politeness in the manners.*

The virtues we are shown here are always less what one owes others than what one owes oneself; they are not so much what calls us to our fellow citizens as what distinguishes us from them.

One judges men's actions here not as good but as fine, not as just but as great; not as reasonable but as extraordinary.

As soon as honor can find something noble here, honor becomes either a judge who makes it legitimate or a sophist who justifies it.

It allows gallantry when gallantry is united with the idea of an attachment of the heart or the idea of conquest; and this is one reason mores are never as pure in monarchies as in republican governments.

It allows deceit when deceit is added to the idea of greatness of spirit or greatness of business, as in politics, whose niceties do not offend it.

… I have said that, in monarchies, educations should bring a certain frankness in the mores. Therefore, truth is desired in speech. But is it for the love of truth? Not at all. It is desired because a man accustomed to speaking the truth appears to be daring and free. Indeed, such a man seems dependent only on things and not on the way another receives them.

This is why commending this kind of frankness here, one scorns that of the people, which has for its aim only truth and simplicity.

Finally, education in monarchies requires a certain politeness in manners. Men, born to live together, are also born to please each other; and, he who does not observe the proprieties offends all those with whom he lives and discredits himself so much that he becomes unable to do any good things.

… Education bears on all these things to make what is called the *honnête homme,* who has all the qualities and all virtues required in this government.

… There is nothing in monarchy that laws, religion, and honor prescribe so much as obedience to the wills of the prince, but this honor dictates to us that the prince should never prescribe an action that dishonours us because it would make us incapable of serving him.

Crillon refused to assassinate the Duke of Guise, but he proposed to Henry III that he engage the duke in battle. After Saint Bartholomew's Day, when Charles IX had sent orders to all the governors to have the Huguenots massacred, the Viscount of Orte, who was in command at Bayonne, wrote to the king, "Sire, I have found among the inhabitants and warriors only good citizens, brave soldiers, and not one executioner; thus they and I together beg Your Majesty to use our arms and our lives for things that can be done." This great and generous courage regarded a cowardly action an impossible thing.

For the nobility, honor prescribes nothing more than serving the prince in war; indeed, this is the preeminent profession because its risks, successes, and even misfortunes lead to greatness. But honor wants the arbiter in imposing this law; and if honor has been offended, it permits or requires one to withdraw to one's home.

It wants one to be able indifferently to aspire to posts or to refuse them. The principal rules are that we are indeed allowed to give importance to our fortune but that we are sovereignly forbidden to give any to our life.

The second is that, when we have once been placed in a rank, we should do or suffer nothing that might show that we consider ourselves inferior to the rank itself.

The third is that, what honor forbids is more rigorously forbidden when the laws do not agree in proscribing it, and that what honor requires is more strongly required when the laws do not require it. (1989, 31–34)

Montesquieu defines his class and its ideas clearly and fully. He does not wish to meld this class with others. Montesquieu seeks to *integrate* aristocratic sovereignty into a governance that is balanced with the realities of will held by a contending "station," the commoners, particularly those of property. He does not seek either to overcome the existing conflict with Republicanism, or sacrifice any of the realities of his traditional system of aristocracy. He writes in the spirit of the fourth phase in seeking a system of in-common governance that finds ways of cooperation. I will address his ideas of a balance of power with the commoners below. Let me say at this point that he, as will Rousseau, be strongly influenced by René-Louis de Voyer de Paulmy, Marquis d'Argenson (1694–1757), who was Minister for Foreign Affairs for Louis XV from November 1744 to January 1747. D'Argenson wrote on the "general will," an idea taken up by Jean-Jacques Rousseau in his *Social Contract* of 1762. The "general will" was not the majority will, rather it was what contending parties could agree upon. Rousseau offers a footnote of how D'Argenson defined the "general will":

"Every interest," says the Marquis d'Argenson, "has its different principles. Harmony between two interests is created by opposition to that of a third." He might have added that the harmony of all interests is created by opposition to those of each. If there were no different interests, we should hardly be conscious of a common interest, as there would be no resistance to it; everything would run easily of its own accord, and politics would cease to be an art.[7]

Rousseau is of the first phase of the next Modernist metaparadigm (c.1755–1865), a decade after D'Argenson wrote. Rousseau will individualize his understanding of D'Argenson. The preceding paragraph to this quote shows Rousseau focus on the individual to a degree that we will not see in the previous fourth phase of when Montesquieu and D'Argenson lived and wrote:

> There is often a great difference between the will of all (what individuals want) and the general will; the general will studies only the common interest while the will of all studies private interest, and is indeed no more than the sum of individual desires. But if we take away from these same wills the pluses and minuses which cancel each other out, the sum of the difference is the general will. (1969, 72–73)

Rousseau speaks of locating the "general will" by what is in-common among individuals once their differences are dropped. The Marquis d'Argenson speaks of differing "stations" and their principles—collective from the onset. To find the collective differences between two "stations" and what is in-common, one studies what I will refer to as the proto-sociological sense of values and principles associated with that "station," as does Montesquieu in his review of what it means to be an aristocrat.

Here one sees what Thomas Kuhn spoke of when he points out that a new, emergent metaparadigm uses the language of the older one in a new way, reconceptualizing the older concept. Rousseau will be the hallmark spokesman for individuality in the initial three phases of the metaparadigm that begins c.1760. Individualization for Montesquieu and D'Argenson has to do with an individual choosing to abide by the collective norms. To find the "general will" in that society is to find principles of each "station" that are in-common.

One can also say that the Marquis d'Argenson was informed by Leibniz, in the earlier phases of the metaparadigm they shared, reconceptualizing the idea of coincident states-of-affairs insofar as parts that conjoin, but each having a section that is non-coincident. That is, in the whole of the society RX, there are differing "stations" composed of "A," "B," and "C" principles. "C" contains the non-coincident principles "A" and "B" have toward each other; but "A" and "B" do have in-common principles within the societal RX as a whole.

---

7. Jean Jacques Rousseau, *The Social Contract*, trans. Maurice Cranston (Harmondsworth, Middlesex, England: Penguin Books, 1969), 73.

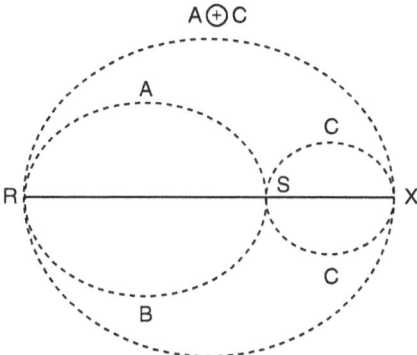

Leibniz focused upon a logical calculus for individual arguments, not proto-sociological realities, where an individual's differences with others of their own "station" or other "stations"
refer to the larger principles that define their membership within their "stations." Rousseau's
thought will move beyond "stations," considering the democratic equality of all individuals,
"stations" thus disappearing.

When Montesquieu then talks of the education of the "commoner," he speaks of what
is appropriate for that "station" if a balance of power is to be achieved that can be
accepted in-common. Commoners, Montesquieu argues, have the best government
when all renounce individual interests that would be averse to societal cooperation. The
laws of a legislature can enforce the range of interactions that subdue unchecked individual passion to have more at the expense of others. Such laws could hamper monarchy
insofar as the differences between the stations of aristocracy, republicanism and even
democracy. Montesquieu refers to the Greek democracy and its possible reoccurrence
which would include even stricter uniformity of self-denial in the interest of curbing the
wealth of individuals. He compares this position with orders of Monks, whose dedica-
tion as painful as it is, provides a model for the uniformity he sees in democracy, which
in itself give a sense of happiness to all who see each other as equally limited, and thus
equal in the fundamental "life world" of everyday existence. The balance of power with
the commoners would allow them to make the laws which could restrict all, including
aristocrats insofar as inequalities of wealth, but with an understanding—if it be a constitutional monarchy—of the final policy privileges of the monarch and his ministers.
Such laws would stress a "moderation" that did not wholly restrict either privileges of not
only the monarch's "station," but the aristocrats who shared that "station." Indeed, this
moderation should lean more to republicanism in its respect for property than the strict
material equality of democracy:

In an aristocracy, if the people are virtuous, they will enjoy almost the same happiness as in
popular government, and the state will become powerful. But, as it is rare to find much virtue
where men's fortunes are so unequal, the laws must end to give, as much as they can, a spirit
of moderation, and they must seek to reestablish the equality necessarily taken away by the
constitution of the state.

The spirit of moderation is what I called virtue in aristocracy; here it takes the place of the spirit of equality in the popular state.

If the pomp and splendor surrounding kings is a part of their power, modest and simplicity of manners are the strength of nobles in an aristocracy. When the nobles affect no distinction, when they blend with the people, dress like them, and share all their pleasures with them, the people forget their own weakness.

Each government has its nature and its principle. The aristocracy must, therefore, not assume the nature and principle of monarchy, which would happen if nobles had any personal and particular prerogatives distinct from the body to which they belong; privileges should be for the senate, and simple respect for the senators. (1989, 51–52)

Montesquieu in his remarks on Monks (1989, 43), as well as his historical examples of individuals who were inconsistent in terms of their behavior and their vaunted principles, demonstrates his own ability to "being aware of how one is aware" in the spirit of this significant development in Modernism. Montesquieu's criticism breathes this hyper-awareness of the differing perspective possible in one individual, indeed as an explanation for why one must be self-aware in this sense. Montesquieu writes, for example, of Solon:

In Athens, Solon acted inconsistently with the old laws, which ordered that goods should remain in the family of the testator, when he permitted on to leave one's goods to whomever one wanted by testament provided one had no children. He acted inconsistently with his own laws; for, by cancelling debts, he had sought equality.

The law that forbade one to have two inheritances was a good law for democracy. It originated in the equal division of the lands and portions given to each citizen. The law did not want any one man to have several portions. (1989, 45)

Montesquieu does not give us the reason for this inconsistency in Solon, but one can infer that Solon was unaware of his own self-violation, or, he was aware of this inconsistency in his perspectives, yet had a self-interest.

B. Literature

The provision, then, which we have here made is no other than Human Nature. Nor do I fear that my sensible reader, though most luxurious in his taste, will start, cavil, or be offended, because I have named but one article. The tortoise—as the alderman of Bristol, well learned in eating, knows by much experience—besides the delicious calipash and calipee, contains many different kinds of food; nor can the learned reader be ignorant, that in human nature, though here collected under one general name, is such prodigious variety, that a cook will have sooner gone through all the several species of animal and vegetable food in the world, than an author will be able to exhaust so extensive a subject.

An objection may perhaps be apprehended from the more delicate, that this dish is too common and vulgar; for what else is the subject of all the romances, novels, plays, and poems, with which the stalls abound? Many exquisite viands might be rejected by the epicure, if it was a sufficient cause for his contemning of them as common and vulgar, that something was to be found in the most paltry alleys under the same name. In reality, true nature is as difficult to be met with in authors, as the Bayonne ham, or Bologna sausage, is to be found in the shops.

But the whole, to continue the same metaphor, consists in the cookery of the author; for, as Mr Pope tells us—

"True wit is nature to advantage drest;

What oft was thought, but ne'er so well exprest."

Henry Fielding, Chapter i. — The introduction to the work, or bill of
fare to the feast. Tom Jones (1749)

There are several characteristics of the fourth phase of thought in the above excerpt that begins Fielding's novel. The most significant is his testimony to "human nature" as the core of the book. We will see in his writing the determinism that gives individuals their manner of thought and action, the human nature which while differing in ways in every person, nonetheless has aspects that condition all of us in our basic responses to one another. The second characteristic is that of the societal norms which indicate how distinctions of "station" give some more privilege than others in their life world, and channel the determined human nature in differing ways given the norms of how one interacts with others in society. There can be "Bayonne ham" for some, and mere ham for others. The third characteristic is "being aware of how one is aware," showing the differences through fiction of those who at times express themselves as "ham," and at other times as "Bayone ham" when engaged in the interpersonal encounter. Fielding will use "metaphor" as above, to show the differences. Pope gives him license to use his reflective wit, metaphorically and with irony or satire to show how individuals "dress their thought" to function within their "station" of society. Fielding does not see anyone as escaping this social determinism, any more than they can escape the human nature which generated the sublimations.

A case in point of even a brilliant author who in Fielding's eyes has fallen prey to certain prejudices that mask human nature in its societal sublimation is that of Samuel Richardson's *Pamela, or Virtue Rewarded* (1740). Richardson sees Pamela as a model for the essential goodness of human nature, and has her, in the spirit of the fourth phase, find a peaceful cooperation with one time enemies, enabling her to find the goodness behind or alongside the evil. After being sexually attacked, abused, kept under lock and key, and sought by the same person to be given away forcibly, Pamela eventually forgives the man, and marries him. Even then, she calls him her "master." The text is epistolary. This is an addition of literary technique, allowing the writer to include a greater context while enabling the heroine who writes the letters to be reflective, while descriptive, re-creating the dialogical interaction she experienced. We will see Johann Wolfgang von Goethe use this technique with quite different aims in the first phase of the next metaparadigm— interesting evidence of what Thomas Kuhn indicated as the reconceptualization of an older concept.

Writing to her parents throughout her ordeals, she gives an account to them of her personal reflections, where she reviews her shifting perspectives toward her former persecutor, shortly before their marriage, Pamela expresses her own faults, with humility and the "forgiveness" of this Mr. B:

> My good dear master, my kind friend, my generous benefactor, my worthy protector, and, oh! all the good words in one, my affectionate husband, that is soon to be—(be curbed, in my proud heart, know theyself, and be conscious of thy unworthiness!)—has just left me, with the kindest, tenderest expressions, and gentlest behavior, that ever blest a happy maiden. He approached me with a sort of reigned in rapture. My Pamela! said he, may I just ask after your employment? Don't let me chide my dear girl this day, however. The two parsons will be here to breakfast with us at nine, and yet you are not a bit dressed! Why this absence of mind, and sweet irresolution? ...

> My God almighty, sir, said I, reward all your goodness to me!—That is all I can say. But, oh! how kind it is in you, to supply the want of the presence and comfortings of a dear mother, of a loving sister, or, of the kind of companions of my own sex, which most maidens have, to soothe their anxieties on the so near approach of so awful a solemnity! You, sir are all these tender relations in one to me! Your condescension and kindness shall, if possible, embolden me to look up to you without that sweet terror, that must confound poor bashful maidens, on such an occasion, when they are surrendered up to a *more* doubtful happiness, and to half-strange men, whose good faith, and good usage of them, must be *less* experienced, and is all involved in the dark bosom of futurity, and only to be proved by the event.

> This, my dear Pamela, said he, is most kindly said! It shows me that you enter gratefully into my intention. (1980, 364–65)

Living in a post-Freudian phase of metaparadigmatic knowledge, I must indicate the depth of sublimation, and, masochistic self-punishment in this exchange, reported by letter to her parents, by Pamela. Freud augmented "being aware of how one is aware" by the level of conscious intention that one rarely recognizes. That Mr. B uses the expression "you enter gratefully into my intention" is almost a comic self-exposure in the post-Freudian mentality.

However, a contemporary to the publication of this novel by Richardson, Henry Fielding, wrote soon after a satire on *Pamela, Or Virtue Rewarded* with his *An Apology for the Life of Mrs. Shamela Andrews*! (1741).

The subtitle of what is now generally referred to in its contemporary editions, as simply *Shamela*, has the irony/satire which was endemic in the early decades of the eighteenth century, most markedly in the works of Jonathan Swift. Irony/satire enables a third phase mind to castigate an enemy, and a fourth phase mind to indicate where a modification of one's contentions can bring a more realistic, cooperative result in actuality. Fielding, in this spirit, adds to his own title:

> In which, the many notorious Falsehoods and Misrepresentations of a Book called Pamela are exposed and refuted; and all the matchless Arts of that young Politician set in a true and just Light. Together with A full Account of all that passed between her and Parson *Arthur Williams*; whose Character is represented in a manner something different from that which he bears in *Pamela*. The whole being exact Copies of authentick Papers delivered to the Editor.

Necessary to be had in all Families. By Mr. Conny Keyber (London: Printed for A. Dodd, at the *Peacock,* without *Temple-bar,* M.DCC/XLI).[8]

Parson Williams will be a potential savior in Richardson's *Pamela,* until he is arrested and put in a debtor's prison by Mr. B for default on a loan. In Fielding's *Shamela* this same incident is noted, but after Parson Williams's release from prison, both he and Pamela will appear to continue an intimacy, hidden from Mr. B (named Mr. Booby by Fielding), after her marriage to the latter. The marriage is depicted by Fielding as solely Pamela's ability to climb in "station" by her seductive engagements with Mr. B, otherwise a loveless marriage.

Essentially, Fielding shows that insofar as sexual appetite, the purity sustained by Richardson's Pamela is not human nature. Recreating a scene in *Pamela* where the servant Mrs. Jencks enables Mr. B to enter a common bed shared with Pamela and her, Mr. B disguised as a woman, and Pamela "naturally" denies this attempt at sex, Fielding offers the following account:

> Mrs. *Jervis* (i.e. Richardson's Mrs. Jencks) and I are just in bed, and the door unlocked; if my master should come—Odsbobs! I hear him just coming in the door. You see I write in the present tense, as Parson *Williams* says. Well, he is in bed between us, we both shamming a sleep, he steals his hand into my bosom, which I, as if in my sleep, press close to me with mine, and then pretend to awake—I no sooner see him, but I scream out to Mrs. *Jervis,* she feigns likewise but just to come to herself; we both begin, she to becall, and I to bescratch very liberally. After having made a pretty free use of my fingers, without any great regard to the parts I attack'd, I counterfeit a swoon. Mrs. *Jervis* then cries out, O, sir, what have you done, you have murthered your poor *Pamela*: she is gone, she is gone.—
>
> *O what a difficulty it is to keep one's countenance, when a violent laugh desires to burst forth.* (1980, 543–44)

The prurience is also in Richardson's account; indeed, although Mr. B's efforts are repulsed by Pamela, the entry to her bed, shared with Mrs. Jencks, by Mr. B disguised as a woman in a shift, is even more designed to engender sexual feeling. Secular realities have entered literature in the sense of sexual nature having certain inescapable causes and effects.

C. Biology—Environmental Determinism as the Cause of the Evolution of Plants, Animals and Homo Sapiens

Montesquieu (1689–1755)

"Evolution" was a concept that existed even before it took on the sense in the mid-eighteenth century by which it is now understood. Montesquieu would not have challenged the existing theory that the male sperm evolved into an embryo warmed by the womb, the woman's only role. However, his views on the role of the natural environment are still current.

---

8.   Henry Fielding, *Shamela* (New York: New American Library, 1980), 529.

Montesquieu both saw the sameness of human nature and differences within the sameness, given the role of climate and geography. He writes as an introduction to his determinative theory:

> If it is true that the character of the spirit and the passions of the heart are extremely different in the various climates, *laws* should be relative in these passions and to the differences in these characters. (*The Spirit of the Laws*, 1989, 231)

He continues, giving not only the effects of climate, but how that in its development of human character, influences history, by conditioning the behavior of typical individuals of that region:

> Cold air contracts the extremities of the body's surface fibers; this increases their spring and favors the return of blood from the extremities to the heart. It shortens these same fibers; therefore, it increases their strength in this way too. Hot air, by contrast, relaxes these extremities of the fibers and lengthens them; therefore, it decreases their strength and their spring.
>
> Therefore, men are more vigorous in cold climates. The action of the heart and the reaction of the extremities of the fibers are in closer accord, the fluids are in a better equilibrium, the blood is pushed harder toward the heart and, reciprocally, the heart has more power. This greater strength should produce many effects; for example, more confidence in oneself, that is, more courage; better knowledge of one's superiority, that is, less desire for revenge; a higher opinion of one's security, that is, more frankness and fewer suspicions, maneuvers, and tricks. Finally, it should make very different characters. Put a man in a hot, enclosed spot, and he will suffer, for the reasons just stated, a great slackening of the heart. If, in the circumstance, one proposes a bold action to him, I believe one will find him little disposed toward it; his present weakness will induce discouragement in his soul; he will fear everything, because he will feel he can do nothing. The people in hot countries are timid like old men; those in cold countries are courageous like young men. If we turn our attention to recent war, such as The War of Spanish Succession, which are the ones we can best observe and in which we can better see certain slight effects that are imperceptible from a distance, we shall certainly feel that the actions of the northern peoples who were sent to southern countries, such as in Spain, were not as fine as the actions of their compatriots who, fighting in their own climate, enjoyed the whole of their courage. (Ibid., 231–32)

Montesquieu then give an account of his own hands-on research that led to these, perhaps, overextended generalizations and conclusions:

> The nerves, which end in the tissue of our skin, are made of a sheaf of nerves. Ordinarily, it is not the whole nerve that moves, but an infinitely small part of it. In hot countries, where the tissue of the skin is relaxed, the end of the nerves are open and exposed to the weakest action of the slightest objects. In cold countries, the tissue of the skin is contracted and the papillae compressed. The little bunches are in a way paralyzed; sensation hardly passes to the brain except when it is extremely strong and is of the entire nerve together. But imagination, taste, sensitivity, and vivacity depend on an infinite number of small sensations.
>
> I have observed the place on the surface tissue of a sheep's tongue which appears to the naked eye to be covered with papillae. Through a microscope, I have seen the tiny hairs, or a kind

of down, on these papillae, between these papillae were pyramids, forming something like little brushes at the ends. It is very likely that these pyramids are the principle organs of taste.

I had half of the tongue frozen, and, with the naked eye I fund the papillae considerably diminished; some of the rows of papillae had even slipped inside their sheaths. I examined the tissue through a microscope; I could no longer see the pyramids. As the tongue thawed, the papillae appeared again to the naked eye, and, under the microscope, the little brushes began to reappear. (Ibid., 232–33)

Montesquieu goes on to characterize the people, and by implication, the history of various societies around the globe. This manner of historical judgment, rooted in biological determinism, will be seen as we go forward in every century in this particular fourth phasal thought. Charles Darwin will give evidence of it between the late 1820s and 1859. Fascist thought in Italy and Germany, in particular, will voice it in the 1920s through the 1940s; although it will continue until the late 1950s in ecological studies.

What is sought in the emergence of science in this first Modern metaparadigm are laws that explain cause and effect in the natural world, without engaging the Divine in final cause or formal cause, and certainly not in efficient. Material cause becomes more primary in the fourth phase because of the need to find commonality with all thinkers. Even as in the first three phases of the metaparadigm, material realities are examined, it is there, as we have seen, that individual intentionality is more primary in the thought of both philosophers of history, writers of fiction, and, even mathematicians who create the logic of calculus, such as Leibniz. Indeed, there is "intentionality" in the study of human character by Locke and Leibniz, the former arguing that the conceptual basis of understanding is a chosen avenue of reality, and for the latter, the impulsion of one's "rational appetite" (Leibniz 1969, 662–63) that enables us to individuate in the proper direction. Choice dominates the third phasal reality, while being determined by material realities dominates the fourth. How people act and judge are the prominent purview, but here in the fourth phase it is how they behave in a pre-intentional manner.

While Montesquieu was perhaps too premature in the implications of the nervous system on the character of historical action, Linnaeus, his contemporary in botany, generated a theory of inclusion for other researchers that was more cautious, yet more thorough, and thus more borrowed by others. Linnaeus (1707–1778) began to publish his taxonomy of all animals, vegetables and minerals, *Systemae Natura* in 1735. Linnaeus had inherited a definitional system from two previous researchers, who had written mainly in the second and third phases of this metaparadigm—Joseph Pitton de Tournefort (1656–1708) and Sebastian Vaillant (1669–1722).[9] In the spirit of what began in the third phase and carried into the fourth, his present, Linnaeus stressed empirical observation, testing the concepts developed gradually in the initial two phases. He wrote in 1735:

In natural science the principles of truth ought to be confirmed by observation. (Ibid., 31)

9. Frans A. Stafeu, *Linnaeus and the Linnaeans. The Spreading of their ideas in systematic botany, 1735–1789* (Utrech: A. Oosthoek, 1971), 62, 119.

Tournefort published his observations, begun in 1686, of the botanical gardens in Paris *Éléments de botanique ou methode pour connaître les plantes* (1694, 1700). Vaillant studied under Tournefort at the botanical gardens in Paris, publishing his taxonomic work *Botanicon Parisiense, ou Denombrement par ordre alphabetique des plantes* in 1727. Linnaeus gives Tournefort credit for being the first botanist to create stricter categories, "families" that separated species. He used the dichotomies as what he termed "synoptic" presentations of the varieties of species and their members (ibid., 62). Linnaeus' observation, guided by dichotomous categories, included five levels of classes "Classes (in a special sense), orders, genera, species, and varieties" (ibid., 61). In the spirit of a metaparadigmatic appreciation of all disciplines in the arts and sciences, he lauded how geography, military science and philosophy used the five categories:

> Examples taken from other sciences make this clear: Geography: Kingdom, province, territory, parish, village. Military: Regiment, company, platoon, squad, soldier. Philosophy: Genus, summum, genus intermedium, genus proxium, species, individuum. (Ibid., 61)

The thinking in this final phase of a metaparadigm is intended to create systematic approaches to problem solving the many fields of inquiry in the arts and the sciences, yet within categories that while clearly differentiated, do not have the fine distinctions that one will see when they are challenged by the initial phases of the next metaparadigm. In the eighteenth century, we now will take up that second Modernist metaparadigm that extends from the late 1750s into the mid-nineteenth century, c.1865.

Nonetheless, new problems were formulated and potentially solved in this first Modern metaparadigm. In philosophy of history, the seminal ideas of every individual being sovereign in himself or herself was introduced by Hobbes, taken further systematically by Locke, and given a temporary solution in the balance of power theories of Bolingbroke and Montesquieu. They were tried, but in the short-run, although the monarchist position never allowed real coincident sovereignty in commoners. However, the cooperative efforts led to new thought in the second metaparadigm that follows. And, by the fourth phase of this second metaparadigm, new cooperative solutions between differing "classes" of society will become sufficiently effective to credit a problem solution by 1865.

The institutional development that encouraged Bolingbroke and Montesquieu was furthered by individuals "becoming aware of how they and others were aware," and this in turn was furthered by the support of breakthroughs in secular literature, as well as the development of a logical calculus that clarified how people reflect and make arguments. One even was able to the difference in how events were structured by individuals in the philosophy of history of Leibniz and Chladenius.

Literature and drama, as well as essays on everyday life, as with Joseph Addison, heightened the new knowledge of people speak with each other, and how certain rhetorical styles are associated with certain norms of civility, as well as they differ among classes or "stations." This gave inquiry into individuality a heightened ability for "being aware of being aware," and, with the contributions of the fourth phase, helped toward solving the problem of a cooperative society.

Montesquieu on human nature as determined by the natural environment created a foundation for explaining behavioral differences among differing nations. This will increasingly be returned to in the spiral development of every fourth phase of the Modernist metaparadigms. Let us say that this problem is not so much solved in terms of mutual understanding, but the bases of it are increasingly understood as social psychology broadens its purview of world cultures. Both the approach of "intentional" choice in the initial three phases of a metaparadigm, and the "determinative" aspects of natural cause are foundations for posing and solving problems of human intercourse in society. Montesquieu's determinative cause is the material environ and the human body, but, as the fourth phase of collective realities, where people seek to solve the problem of peaceful cooperation, this conceptual determinism, insofar it is a societal set of norms, will be an added helpmeet. Indeed, the work of literary minds such as Richardson and Fielding give us these normative societal insights into human choice, and in that a pre-sociology that in future Modern metaparadigms will improve how we formulate and solve societal problems.

Part II

The Second Modern Metaparadigm, c.1750–c.1865

1745 E. Young *Consolation* 97 Nature delights in Progress; in Advance From Worse to Better: But, when Minds ascend, Progress, in Part, depends upon Themselves.

Edward Young, Oxford English Dictionary

This metaparadigm will see the onset of the concept of "progress" in the inquiry and findings of all the arts and sciences. Moreover, it will stress the individual challenge and leadership in this endeavor in the initial three phases. The word "progress" did exist since the Early Enlightenment, but mostly in the understanding of advancement, but not in the concept of Progress as the rationale for human problem-solving that became the hallmark of the Modernist era. Young is the first who wrote on this subject extensively.

Additional epigraphs of the age:

This author stands now upon a hill where he can see more than the limited way others see before them (Prologue). What the education of single individual is, is the revelation for all humankind (Par. 1). Education is revelation that occurs for the individual; and revelation is education that has occurred for humankind, and still can occur (Par. 2) …

Education gives the person nothing that he cannot have on his own, only it is quicker and easier. Indeed, revelation cannot give humankind anything they cannot themselves grasp from their own reason, if left to themselves, only it gave and gives him the most important of things earlier (Par. 4).

There can be a third spirit of the times (Par. 89), beyond the lowest level of spiritual knowledge we received as humankind in our youngest years, when the old testament gave us law (Pars. 26 and 27). Beyond that as adolescent humankind we received the higher level of truth where the eternal nature of Christ's spirit within us pointed our way (Par. 75). Now, we can ourselves be the truth sayer based upon our own self-understanding, and the understanding of others (Par. 86).

Gottfried Ephraim Lessing, The Education of Humankind (1780)

Lessing has captured in this observation the spirit of the Late Enlightenment, as well as the decades that follow in this Second Modern metaparadigm. Humankind is now self-aware, and learning more about its own complexity. Indeed, the world can be transformed by humankind itself, setting its own problematic goals and devising the means to realize them. If there is a Divine, and most hold that there is, we are so fashioned to discover the truth of existence ourselves, with our own intentions, with our own interventions methodically into what is around us, so as to improve our existence. There is only human progress, and the solving of every problem we face with our own devices.

Chapter Five

THE FIRST PHASE: SEMINAL IDEATION, C.1750–C.1770: THE FOCUS UPON DEFINITION AND HYPOTHESIS

Look down here, Zeus,

Upon my world: it lives,

And I have formed it after my own image,

A race that shall resemble me,

To grieve and weep, to enjoy and to delight, And heed you as little as I.

Johan Wolfgang von Goethe, Prometheus (c.1772)

Just as in the vegetable and animal kingdoms an individual begins, so to speak, grows, continues to exist, degenerates, and is no more, so it might well be with species in their entirety. If faith did not teach us that animals spring from the hands of their Creator just as we now see them and if it were permissible to entertain the slightest doubts about their beginning and their end, might the philosopher not suspect, having given himself up entirely to his own conjectures, that the particular elements needed to constitute animal life had existed from all eternity, scattered and mixed in with the whole mass of matter; that these elements, happening to come together, had combined because it was possible for them to do so; that the embryo formed by these elements passed through an infinite number of structural changes and developments; that it acquired, successively, motion, sensation, ideas, thought, reflection, conscience, feelings, passions, signs, gestures, sounds, articulated sounds, a language, laws, sciences, and arts; that millions of years elapsed between each of these developments; that, unknown to us, it may have further developments still to undergo and further accretions to acquire; that it has experienced, or will experience, a period of stability; that it is moving out of, or will move out of, that period into a long decline, during which it will lose its faculties just as it acquired them; that it will disappear from nature forever or, rather, that it will continue to exist in nature but in another form and with faculties quite other than those we observe in it at the present moment of its duration? Religion spares us many wanderings and much labor. If it had not enlightened us as to the origin of the world and universal order that governs phenomena, think how many different hypotheses we should have been tempted to accept as the secret of nature. And those hypotheses, since they are all equally false, would all have seemed to us more or less equally probable. The question *why anything exists* is the most embarrassing that philosophy could ever have asked itself: only revelation can answer it.

Diderot, On the Interpretation of Nature (1753)

Goethe's challenge to the Divine of human independence, and Diderot's vision of the complex development of humankind in its own conceptual creations that have increasingly furthered its abilities, articulate how it is "to be aware of how one is aware," generating even more awareness of how one thinks and acts. This is the spirit of the Late Enlightenment, which can be dated from the onset of the second Modern metaparadigm, 1755, through the revolutionary years of France. The Seven Years War marked the final transition to the Late Enlightenment onset of contention with tradition that bred an individualism that was absolute, and a willingness of each individual to stand alone, possessing powers that were conflated with the Divine, but now owned by the Western intellectual. From deism, to agnosticism, or even atheism, the individual broke free in a manner that did not look back toward any other final cause but him- or herself. Feeling the self-created power of the First Industrial Revolution, after the steam engine of John Watts that could power factories 24 hours a day, that eventually created engines that were faster than any animate creature, no hesitation to consider himself or herself capable of transforming the planet. The new theory of human evolution, women's coequal power in reproduction, and abilities to hold their own in any occupation, created a new contention between the sexes that transformed the meaning of human partnership. We will see in this chapter in the philosophy as well as practice of history, insofar as institutional development, in literature, and in biology, the creation of new ideas of being human that went further than any earlier human culture.

A. Philosophy of History/Development of Institutions

Jean-Jacques Rousseau (1712–1778)

> My purpose is to consider if, in political society, there can be any legitimate and sure principle
> of government, taking men as they are and the laws as they might be.
>
> Rousseau, "Initial sentence" of Book I, *The Social Contract* (1762)

Rousseau's purpose is to put forth differing concepts of human nature and human association, in order, to prepare for a more systematic plan for developing a feasible, enduring democracy that frees individuals to express their mind and will. It is based upon his understanding of human nature, but unlike the generalized sense of a determined human nature that is unchangeable, or natural conditions that create the character of individuals in certain geographies, his view is of "intentional" individuals who can in any clime or existing system, find their individual voices. In this he is much like John Locke, who wrote in the third phase of the first Modern metaparadigm of the human mind as a *tabula rasa* that a new education could prepare for a newer, free horizon of chosen action. One might say that Rousseau took up the very radical sense of Hobbes that human autonomy meant one could never stop fighting for a personal individual freedom and autonomy, even if that meant breaking free of any form of incarceration (*Leviathan*, 1967, 170 [Art 2, Chap. 21]). One can call this vision of Rousseau that echoed Hobbes' first phasal *Leviathan* a "spiral return" to the principle, but one that augmented it. For Hobbes, the exercise of absolute freedom was considered when an individual was hampered in exercising his or her normal, rational being.

Rousseau placed an interesting limit on that which was informed by the new proto-sociology he experienced in the writing of Montesquieu, Bolingbroke, Richardson and Fielding—the existence of social "stations" with differing norms. Rousseau's concept of the "general will," discussed earlier in my treatment of Montesquieu, where Rousseau will create an interesting use of the Marquis d'Argenson's concept of the "general will." Montesquieu and the Marquis d'Argenson understood an invariant difference between the "stations" of aristocracy and commoner. Rousseau will have a proto-sociological awareness of that, but will refuse to see anything but an equality of all, based on being human. Rousseau, of course, is writing just after the emergence of evidence of such biological equality in the work of the first evolutionists, Friedrich Caspar Wolff (1733–1794) in biology and Joseph Kolreuter (1733–1806) in botany. Rousseau did see that there must be an area of in-common understanding that formed public governance, and that the "general will" once formed, must be obeyed by the citizen of the political reality (1969, 69–72). Yet, one's autonomy was absolute, privately, while publically one made a "social contract" with the "general will," no longer put forward as invariant classes of people. The idea of a public duty to the "general will," and a private right that is autonomous, we will see in Immanuel Kant's essay "What is the Enlightenment." Rousseau was a guiding light for Kant's vision of history and institutional development. Indicative of that, besides Kant's writings that will be taken up in the third phase of this second Modern metaparadigm, is the biographical fact that prominent in his living quarters—called by Ernst Cassirer its only decorative element—was a large portrait of Rousseau.[1]

Rousseau's proto-sociology was a new idea, and idea that surfaced in multiple areas of European history as the Seven Years War came to an end in 1763. One thinks of how Lord William Pitt, The Great Commoner, as the Prime Minister for the new monarch George III, said he would resign rather than give the fishing rights off the Gulf of St. Lawrence back to the French, after so many English soldiers died in taking Montreal, ending the war in their favor. George III, nonetheless, gave the rights back to his "cousin" Louis XIII. Pitt's adamant stance was the beginning of a call to revolution. The young, future revolutionary Americans read the works of John Wilkes, whose newspaper defended Pitt.

Rousseau approached this idea of potential revolution in a deconstructive manner of ideation, giving evidence in several writings in 1762 that the individual can elevate himself to a knowledge that made him spokesperson of human equality. This was not mere sloganeering. Rousseau read and thought about the functioning of consciousness,

---

1. Ernst Cassirer, *Kant's Life and Thought*, trans. James Haden (New Haven: Yale University, 1981), 362.

as Thomas Hobbes and John Locke,[2] who wrote in the initial phases, as did he, of the previous metaparadigm. By studying history, examples of past governance, and, the classical philosophers, one's mind can be sufficiently matured to comprehend the key concept of his vision—that is the "general will," and why is "democracy" its best expression in governance. Rather than a collective vision of what can be a flawless governmental structure, even a democratic government, he sees faults in all forms. The necessary corrective is always the individual of understanding. Kant in his later *Universal History from a Cosmopolitan Pont-of-View* (1784), reflecting Rousseau throughout the essay, bemoans that the human is an imperfect creature, made of "crooked wood."[3] This, too, is Rousseau's general vision insofar as one person expressing the sovereign idea of the "general will" as a form of governance. Rousseau prefers "the many individuals" who each must discern what and why their view is in challenging or defending their conception of the "general will" (Rousseau, *The Social Contract*, 149–51 [Book IV, Chapter One: That the General Will is Indestructible]):

> "I might say a great deal here about the simple right of voting in every act of sovereignty, a right of which nothing can deprive citizens, and also about the right of speaking, proposing, dividing and debating." (Ibid., 151)

Rousseau's first sentence in his first chapter is the memorable "Man was born free, and he is everywhere in chains (ibid., 49). Here again one sees his proto-sociology. The individual is autonomous, free to choose his or her path, but the individual as he writes is hampered by his or her "station" in society. Rousseau' high regard for the Marquis d'Argenson overlooks the fact that the Marquis argues for a "general will" that will accommodate the "stations" of society that exist as he works for the government of Louis XIII. The Marquis and Montesquieu, writing in the fourth phase, both sought a balance of power. Yet, this view was a collective one where the social whole dictated certain principled freedoms, but did not dwell on the number and nuances of differences in individual thought. Rousseau ends *The Social Contract* with a quote from the Marquis d'Argenson, complimenting his thought toward the individuality Rousseau culls from it:

> "In the republic," says the Marquis d'Argenson, "everyone is perfectly free to do what does not injure others." Here is the invariable boundary; one could not express it more exactly. (Ibid., 185)

The proto-sociology, insofar as Rousseau speaks briefly of forms of authoritative government, democracy being the best for a flexible and totalizing "general will," challenges

---

2. Rousseau has what the Germans would call an "Auseinandersetzung" with Locke throughout his book on education and the human mind in the person's several stages of maturation from childhood. See Rousseau, *Emile or Education*, trans. Allan Bloom (Basic Books, 1979), 5, 13, 15, 18, 25, 33, 55, 89, 90, 103, 117, 126, 128, 197, 255, 279, 357, 481 n 4, 484 n 21, 485 n 37 and 485 n 47.
3. Immanuel Kant, "Universal History from a Cosmopolitan Point-of-View," in *Kant, On History*, trans. Lewis White Beck (New York: Bobbs-Merrill, 1963), 17–18.

then the existing system sufficiently to be a call for revolution—as it was for the French by the third phase of this metaparadigm. Rousseau's work has a subtitle that indicates its definitional character—quite similar in its several paragraph style, as with Hobbes, in articulating an unfamiliar concept or, as he calls it, "principle." The subtitle is "or Principles of Political Right."

Finally, Rousseau's indebtedness to the previous phase of thought is found in his chapter on the role of environment in determining how individuals think, and why that is significant in their forming a certain kind of government (ibid., 124–29 [Book III, Chapter Eight]). His indebtedness to Montesquieu is quite clear; yet, as he writes he augments toward individuality. He ponders how individuals think and behavior individuals in those regions, and what are the constraints of how might form a functioning "general will." This, of course, is how a fourth phase in its insights must continue in the succeeding initial phases of the next metaparadigm, where individuality and new avenues of thought insofar as a focus upon individual consciousness and behavior are sought. What Book III, Chapter Eight, presents is the foundation for a nationally inflected, self-conscious "historiography." The environmental character of that region, and any defense or attack on a governmental structure of a particular region will reflect the constraints and advantages of that region. The "intentionality" of those responsible for the region's institutional development will have a perspective shaped by that region's influences and necessities. Thus, any political critic or historian of that region must be examined within the formal causal variables of his or her education—both the formal and the informal sociological of his or her milieu. Rousseau in this chapter raises an interesting aspect of "being aware of how one is aware." This historiographical bias has been examined by some historiographers over the years, although not as often or as in as much depth as might be done. Of this type of research, we can see recently in Austria, for example, Alphonse Lhotsky's *Österreichische Historiographie*.[4]

B. Literature

Gotthold Ephraim Lessing (1729–1781) and Edward Young (1683–1765)

Lessing turned stagecraft on its head in that he created multiple scenes within an act in his drama. Rather than Voltaire's "three unities" of time, place and manner, the dramas of Lessing saw each scene, in each act, as having its own unity, the next scene a differing unity, all in the service of exploring the psychological complexity of an individual's emotion, thought and behavior. In Lessing's *Minna von Barnhelm* (1767) there are 56 scene changes in the normatively traditional five acts. In *Emilia Galotti* (1772) there are 43 scene changes in the five acts. Voltaire in his *Merope* (1743) had 30 scenes in five acts; in his *Orestes* (1750), he had 35 in five acts. Lessing used many of his scenes for one person monologues. In *Minna von Barnhelm* there were six. In *Emilia Galotti* there were five. Each monologue articulated the intentions of the character in an appropriate emotion, using a

---

4. Alphonse Lhotsky, *Österreichische Historiographie* (Vienna: Verlag für Geschichte und Politik, 1962).

literary device termed aposiopesis, a grammatical device that becomes normative in this phase—incomplete sentences that show the moment to moment thoughts and feelings of the individual, heightening his or her presence and plight. Voltaire, on the other hand, uses well-articulated, complete sentences, where we are "told" by the isolate character his or her state-of-mind. In *Merope*, Voltaire has three such scenes, and in *Orestes*, three such scenes. The fourth phase of a metaparadigm, which Voltaire commanded in literature, has the aim and effort to integrate with others, counseling subdued emotions. Whereas, the initial three phases of any metaparadigm, as it introduces new avenues of an individual "being aware of being aware" will seek, as did Anton Ulrich and as we will see with Lessing, to bring us closer to the individual in the midst of their struggle to be self-aware, and/or to realize their intention.

Lessing's criticism of Johann Joachim Winckelmann's *Reflections on the Imitation of Greek Works in Painting and Sculpture* (1755) captures the essential spirit of this first phase of the second Modernist metaparadigm. One sees in the several paragraphs that begin the text of his 1767 *Laocoön, An Essay on the Limits of Painting and Poetry* an expression of the conceptual differences that determine human thought and behavior from the last phase, that of a sense of collective determinism, toward which individual reality could only adapt, reaching compromise rather than new beginnings. The focus will again be upon intentionality, a quantum change in interpretation from the previous fourth phasal perspective among the disciplines, and, rather than a determinism that has to be accommodated, a challenge to the affects of moment to moment realty, an effort to contest and shape it. Each individual will need to be closely studied in his or her "wrestling" with the moment. I quote at length here, for the "auseindersetzung" with Winckelmann is telling:

The general and distinguishing characteristics of the Greek masterpieces of painting and sculpture are, according to Herr Winckelmann, noble simplicity and quiet grandeur, both to posture and expression. "As the depths of the sea always remains calm," he says "however much the surface may be agitated, so does the expression in the figures of the Greeks reveal a great and composed soul in the midst of passion."

"Such a soul is depicted in Laocoön's face—and not only in his face—under the most violent suffering. The pain is revealed in every muscle and sinew of his body, and one can almost feel it oneself in the painful contraction of the abdomen without looking at the face or other parts of the body at all. However, this pain expresses itself without any sign of rage, either in his face or his posture. He does not raise his voice in a terrible scream, which Virgil describes his Laocoön as doing (Virgil, *Aeneid* I, 222); the way in which his mouth is open does not permit it. Rather he emits the anxious and subdued sight described by Sadolet. The pain of the body and the nobility of soul are distributed and weighed out, as it were, over the entire figure with equal intensity. Laocoön suffers, but he suffers like the Philoctetes pf Sophocles; his anguish pierces our very soul, but at the same time we wish that we were able to endure suffering as well as this great man does.

Expressing so noble a soul goes far beyond the formation of a beautiful body. This artist must have felt within himself that strength of spirit which he imparted to his marble. In Greece artists and philosophers were united in one person, and there was more than one Metrodorus. Philosophy extended its hand to art and breathed into its figures more than common souls ..."

The remark on which the forgoing comments are based, namely that the pain in Laocoön's face is not expressed with the same intensity that its violence would lead us to expect, is perfectly correct. It is also indisputable that this very point shows truly the wisdom of the artist. Only the ill-informed observer would judge that the artist had fallen short of nature and had not attained the true pathos of suffering.

But as to the reasons on which Herr Winckelmann bases this wisdom, and the universality of the rule which he derives from it, I venture to be of a different opinion.

I must confess that the disparaging reference to Virgil was the first cause of my doubts, and the second was the comparison with Philoctetes. I shall proceed from this point and record my thoughts as they developed in me.

"Laocoön suffers like Philoctetes of Sophocles." But how does Philoctetes suffer? It is strange that his suffering should have left such different impressions. The laments, the cries, the wild curses with their anguish filled the camp and interrupted all the sacrifices and sacred rites resounded no less terribly throughout the desert island, and it was this that brought about his banishment there. What sounds of despondency, of sorrow and despair in the poet's presentation ring through the theater! It has been found that the third act of his work is much shorter than the others. From this, the critic's claim, we may conclude the ancients were little concerned with having acts of equal length. I agree with this, but I should prefer to rely on some other example than this for support. The cries of anguish, the moaning, the disjointed â â, øeû, ἀπαγαῖ, ὠγοί γοί! the whole lines of παπâ παπâ of which this act consists and which must be spoken with prolonged stresses and with pauses quite different from those of connected speech, have in actual performance doubtless made this act just about as long as the others. It seems much shorter on paper to the reader than it probably did to a theater audience.[5]

Note how Lessing in this last paragraph indicates to the reader that the actual cries, pauses and prolonged stresses, what I have seen as aposiopesis, is dominant so that one

---

5. Gotthold Ephraim Lessing, *Laocoön, An Essay on the Limits of Painting and Poetry*, trans. Edward Allen McCormick (Baltimore: Johns Hopkins University Press, 1962), 7–8.

can make closer contact with the individual in his or her moment to moment struggle with existence. The experimentation with forms of dialogue, both poetic and prose, marked the initial phases of the second Modernist metaparadigm. Friedrich Gottlieb Klopstock (1724–1803) was close to Lessing in his spirit of change, of innovation toward the greater presence of the person in his art. Klopstock created over 70 separate forms of poetic diction in his search for presenting "presence" in the lives of his persona. The *Deutsche Biographie* writes in this vein of Klopstock's creative efforts:

> Klopstock distanced himself from the poetic tradition that had developed since Opitz and withdrew from any use of alternating meter, which curtailed the free movement of speech and expression. He eschewed rhyme as well or any melodious schemata. He sought, rather, ancient meters and enriched them. In their model he developed further his own rhyme-less forms of strophe for speech and expression. Tense rhythmic passages, and artful syntactic uses to create differing moments of expression, lent his diction an energy that had been lost under the orthodox use of rhyme. (https://www.deutsche-biographie.de/sfz42987.html?language=en)

Montesquieu contributed to the same article "On Taste" in Diderot's *Encyclopedia* as Voltaire. This is significant in this section on Literature in the first phase of the second Modern metaparadigm because Montesquieu wrote this section of the essay in 1753, two years before he died. This was a time of transition from the fourth phase sense of traditional thought and writing that integrated difference, seeking an in-common style, to the new challenging, person-centered effort to feel into the moment to moment existence. Montesquieu reflects in the essay both the fourth and new first phase. He champions "symmetry" in prose and poetry, but also says:

> Wherever symmetry is useful to the soul and can further its functions, the soul finds it pleasant; but wherever it is useless, it is tedious because it destroys variety. (Diderot, "On Taste," 349)

He goes on with a few paragraphs titled "*Concerning Contrasts*":

> The soul loves symmetry, but it also loves contrasts. This requires a great deal of explanation. For instance: While nature requires painters and sculptors to introduce symmetry into the various parts of their figures, it demands on the other hand that they introduce contrasts into the attitudes of these figures. A foot placed in the same position as another, a limb that moves like another, these are unbearable ... Therefore, figures must be presented in contrasting attitudes, especially in sculpture, which is by nature cold; sculpture can only express the fire of life through strong contrasts, and a striking position. (Ibid., 350)

Montesquieu then addresses "that certain something" that attracts, surprises and pleases. The section is titled "*Concerning the 'je ne sais quoi*'":

> Persons or objects sometimes have an invisible charm, a natural grace which defies definition and perforce has been called *je ne sais quoi*. It seems to me that this effect is chiefly based on surprise. Our feelings are roused because a woman attracts us more than we had expected at

first, and we are pleasantly surprised because she has been able to overcome defects which our eyes reveal to us but which our heart no longer believes to be true. (Ibid., 355)

Montesquieu's "being aware of how he has been aware" is evident for the reader, so that the reader can reflect on this in himself or herself. Montesquieu goes on to say of how one is "charmed" by this surprise. And then tells us the "charm" is in the mind, not a natural feature of the face that charms us. In this, he has shifted from natural determinism to intentional concept, making the move of all first phase thinkers:

> Charm usually resides in the mind rather than the face, for a beautiful face immediately reveals its beauty and scarcely conceals anything. But the mind reveals itself little by little, only when it wishes and as much as it wishes … Charm is found less in the features of the face than in behavior, for behavior varies with every moment and can at any time create surprise. In a word, a woman can be beautiful in only one way, but she is attractive in a hundred thousand ways. (Ibid., 356)

Montesquieu tells us to attend every moment of a person's shifting persona, in the spirit of the new, first phase thought that emerges in the mid-1750s. Did he know that he was articulating a perspective now shared by others, such as Lessing and Rousseau, and as we will see, another contributor to the essay "On Taste," D'Alembert? No, "becoming aware of how one is aware" in the sense of knowing generational change, will await German philosophers and historians, and the coinage of the concept of "Zeitgeist." Indeed, Kant's observation in his *Critique of Pure Reason* (1784) that a thinker will often not know how he or she has formulated an idea, but offers it in a way that we can comprehend it in a moral clear manner than that thinker who articulates it. We, the listeners, are either as one who has achieved "being aware of how one is aware" in that sphere, or as a later mind schooled in the progress of idea this thinker engaged in, albeit not knowingly, can comprehend it. Kant writes:

> I need only remark that it is by no means unusual, upon comparing the thoughts which an author has expressed in regard to his subject, whether in ordinary conversation or in writing, to find that we understand him better than he has understood himself. As he has not sufficiently determined his concept, he has sometimes spoken, or even thought, in opposition to his own intention. (*Critique of Pure Reason*, 1965, 310 [A 314, B 371])

Jean la Rond D'Alembert (1717–1783) writes the final section of "On Taste" in 1757. He urges new inquiry into the immediacy of "taste." He attacks natural determinism and convention:

> The true philosopher judges the pleasure that poetry gives us neither from nature nor from convention. He recognizes that just as music has a universal effect on all nations, even though the music of one nation does not always sound pleasant to another, so too every nation appreciates every nation appreciates poetic harmony, even though the poetry of every nation differs. By examining this difference attentively, the philosopher will succeed in determining to what degree habit influences the pleasure we derive from poetry and music. (Ibid., 366)

D'Alembert and Diderot are at this point at the onset of the new metaparadigm and its first phase that seeks to enlarge "upon what we should be aware, as a new view of how we are aware." "On Taste" is a definitional essay, where concepts are briefly introduced, but no lengthy study is made.

Interestingly, D'Alembert tells us to become aware of differences in thought and creative work between nations, thus providing the lead for Rousseau's thought in the *Social Contract* on this subject, as well as possibly providing insights by Lessing on this same subject. Lessing writes an entire text in 1767 on differences in the theater between nations. In his *Hamburgische Dramaturgy* he focuses upon the English, the French, the Italian and the German theaters, offering conceptual overviews of how they are written and staged. He brings the same acuity of focus upon how individuals are examined in the use of scenes as he had in his own works in that year in the play *Minna von Barnhelm* and his book on the aesthetics of painting and poetry, *Laokoön*.[6] These 1767 texts are definitional, introducing new conceptual lines of inquiry and explanation, without the extended systematic perspective we will see in works in the second phase that begins c.1770.

Yet, the fuller development of definition into concepts that will guide new inquiry into human thought, passion and behavior in this second Modern metaparadigm had an earlier articulation as mere idea in the general sense. That came from Edward Young who in 1745 spoke of "progress" in human understanding by the individual, and the implied lesson that each of us is responsible for having new, greater insight. In his 1759 *Conjectures on Original Composition*[7] he speaks still with a general view of individual originality and its necessity. He challenges how tradition dirempts individual initiative into probing beyond what is known. This attitude and the essay undoubtedly stimulated and reinforced thought across the Western cultures. Young was well-known and read. The young Goethe, for example, in 1766 said that he was learning English by reading Milton and Young.[8] Young denoted the individual who looked in a new manner into human thought and behavior an "original" and also a "genius." What prevents us from being an original? Our sense of sufficiency in knowing the ancients and tradition, and simply building on their avenues of thought. Young writes in the 1759 essay:

> But why are originals so few? Not because the writer's harvest is over, the great repairs of antiquity having left nothing to be gleaned after them; nor because the human mind's teeming time is past, or because it is incapable of putting forth unprecedented births; but because illustrious examples engross, prejudice, and intimidate. They engross our attention, and so prevent a due inspection of ourselves; they prejudice our judgment in favor of their abilities, and so lessen the sense of our own; and they intimidate us with the splendor of their renown, and thus under diffidence bury our strength. Nature's impossibilities, and those of diffidence lie Wide asunder. (Young N.D., 340)

---

6. Gotthold Ephraim Lessing, *Hamburgische Dramaturgie* (Frankfurt am Main: Insel Verlag, 1986).
7. Edward Young, *Conjectures on Original Composition* (Manchester: University of Manchester Press, N.D.).
8. https://en.wikipedia.org/wiki/Edward_Young

Young does not write off tradition, rather he counsels others to find the originality in themselves that great thinkers of antiquity found in themselves. Moreover, one can enter the avenues of thought of a Homer or Aristotle, but in the spirit of going beyond what was seen in that very method of seeing by the selected ancient thinker:

> Must we then, you say, not Imitate ancient authors? Imitate them, bv all means; but imi-tate alright. He that imitates the divine Iliad, does not imitate Homer; but he who takes the same method, which Homer took, for arriving at a capacity of accomplishing a work so great. Tread in his steps to the sole fountain of immortality; drink where he drank, at the true Helicon, that IS, at the breast of nature: imitate; but Imitate not the composition, but the man. For may not this paradox pass into a maxim? viz. "The less we copy the rcnowned ancients, we shall resemble them the more" …

> By the praise of genius we detract not from learning; we detract not from the value of gold, by saying that diamond has greater still. He who disregards learning, shows that he wants its aid; and he that overvalues it, shows that its aid has done him harm. Overvalued indeed it cannot be, if genius, as to composition, is valued more. Learning we thank, genius we revere; that gives us pleasure, this gives us rapture; that informs, this inspires; and is itself inspired; for genius is from heaven, learning from man: this sets us above the low, and illiterate; that, above the learned, and polite. Learning is borrowed knowledge; genius is knowledge innate, and quite our own. (Young 1966, 343)

C. Biology and Botany—Discovery of the Male-Female Participation in the Evolution of the Embryo

Friedrich Caspar Wolff (1733–1794)

> The plan of my work forces me to pass over full explanations of things that contribute to growth, other than presenting an accessible, i.e. conventional order of the parts and how they evolve and disappear.[9]

> Caspar Friedrich Wolff, *Theoria Generationis* (1759)

The development of an embryo in an egg could, indeed, be studied with the technical means available in the eighteenth century. What could be observed there concerned shape and move-ment only, since the question was to distinguish between the development of a preformed germ and the progressive elaboration of the young animal by "epigenesis". By observing the development of the chicken under the microscope, Caspar Frederick Wolff discerned superposed membranes, first simple, then folded, forming the swellings, grooves and tubes from which emerged the rough outlines of organs: the nervous system, then the vessels, the digestive tube, etc. The primary structure of a living being was not, therefore, preformed in the egg. It was gradually organized by a series of pleats, swellings and blisters, a sequence of mechanical operations in time and space.[10]

> Francois Jacob, *The Logic of Life*

---

9. Friedrich Caspar Wolff, *Theoria Generationis* (1759), Part I, Vorrede, Erklärung des plans, Entwicklung der Pflaznen, trans. Dr. Paul Samassa (Leipzig: Wilhelm Engelmann, 1896), 38.
10. Francois Jacob, *The Logic of Life, A History of Heredity,* trans. Betty E. Spillman (New York: Pantheon Books, 1982), 66.

Wolff changed the manner in which the development of life was observed. Moreover, he himself was imbued with a conceptual attitude that sought to see "development" rather than what might be considered a finished growth. Wolff's fellow scientist in the study of plants, Joseph Kolreuter, had new evidence on the changes the character of generational change in plant life (see Jacob 1982, 69). This sense of a constantly changing reality of life fostered a rise in the experiments with hybrids, tracking how they could return to earlier ancestors, but also how they could generate more vital specimens with care in the management of the hybridization by the human agent.

This new knowledge worked against the millennial old understanding that the male seed in organic (animal and human animal) and inorganic nature had a "pre-formed" succession of development simply facilitated by the female. This co-equal participation changed how women were viewed, and gave all scientists a challenge to observe differently than had always been the case. "Evolution" had been the name of this male unfolding, so that now "Evolution" became defined differently.

"Epigenesis," the generation of the living organism from "embryo" to visible specimen, or, animal and human animal infant at birth was now a co-existent male-female generation. The analogs for this entered literature insofar as male-female agency, much as Einstein's theory of relativity entered the language of ordinary and literary thought in the twentieth century. Indeed, insofar as history, the idea of Progress in episodes will be especially evident by the second phase of this Second Modernist Metaparadigm, c.1770–1790. What we call now the "Zeitgeist" was coined as a germ by German historians, many of whom we will see did argue for "Progress" in succeeding "Zeitgeists."

Diderot and D'Alembert, writing to each other about nature as early as 1753, saw a constant change in all species of existence. Diderot expresses this spirit of this sense of newness in knowing existence:

Just as in the vegetable and animal kingdoms an individual begins, so to speak, grows, continues to exist, degenerates and is no more, so it might well be with species in their entirety. If faith did not teach us that animals sprang from the hands of their Creator just as we now see them and if it were permissible to entertain the slightest doubts about their beginning and their end, might the philosopher not suspect, having given himself up entirely to his own conjectures, that the particular elements needed to constitute animal life had existed from all eternity, scattered an mixed in with the whole mass of matter; that these elements, happening to come together, had combined because it was possible for them to do so; that the embryo formed by these elements passed through an infinite number of structural changes and developments; that it acquired successively, motion, sensation, ideas, thought, reflection, conscience, feelings, passions, signs, gestures, sounds, articulated sounds, a language, laws, sciences, and arts; that millions of years elapsed between each of these developments; that, unknown to us, it may have further developments still to undergo and further accretions to acquire; that it has experienced, or will experience, a period of stability; that it is moving out of, or will move out of, that period into a long decline, during which it will lose its faculties just as it acquired them; that it will disappear from nature forever or, rather, that it will continue to exist in nature but in another form and with faculties quite other than those we observe in it at the present moment of its duration? Religion spares us many wanderings and much labor. If it had not enlightened us as to the origin of the world and the universal order that governs phenomena, think how many different hypotheses we should have been tempted to accept as the secret of nature. And those hypotheses, since they are all equally false, would all have seemed to us more or less equally probably. The question *why anything exists* is the most embarrassing that philosophy could ever have asked itself; only revelation can answer it.[11]

I feel safe in intimating Diderot's sarcasm toward Christian revelation. Rather one sees in his deliberation the willingness to genuinely ask, observe and search for an answer, as Wolff was actually doing at this same time. All Western minds that shared the Zeitgeist of this initial phase in the second Modern metaparadigm were willing to reach into the unknown with hypotheses that could address the fundamental issue of existence, and, how it has developed. An English pastor by the name of Laurence Sterne (1713–1768) wrote in 1759 the comic novel *Tristram Shandy* that had as a premise how the thought of one's parents in the moment of intercourse could affect how one developed from the ensuing embryo to adulthood. The agency of the individual parents captured the focus upon individual choice of the emergent metaparadigm in its initial phases. Indeed, one can see a presage of Lamarck's view of evolution written in the third phase. It is worth a lengthy quotation from this text to display the "spirit of the time" in creative thinking:

Chapter I

I wish either my father or my mother, or indeed both of them, as they were in duty both equally bound to it, had minded what they were about when they begot me; had they duly consider'd how much depended upon what they were then doing;—that not only the production of a rational Being was concern'd in it, but that possibly the happy formation and temperature of his body, perhaps his genius and the very cast of his mind;—and, for aught they knew to the contrary, even the fortunes of his whole house might take their turn from the humours and dispositions which were then uppermost.—Had they duly weighed and considered all this, and proceeded accordingly,—I am verily persuaded I should have made a quite different figure in the world, from that, in which the reader is likely to see me.—Believe me, good folks, this is not so inconsiderable a thing as many of you may think it,—you have all, I dare say, heard of the animal spirits, as how they are transfused from father to son, &c. &c.—and a great deal to that purpose:—Well, you may take my word, that nine parts in ten of a man's sense or his nonsense, his successes and miscarriages in this world depend upon their motions and activity, and the different tracks and trains you put them into; so that when they are once set a-going, whether right or wrong, 'tis not a halfpenny matter,—away they go cluttering like hey-go-mad; and by treading the same steps over and over again, they presently make a road of it, as plain and as smooth as a garden walk, which, when they are once used to, the Devil himself sometimes shall not be able to drive them off it.

Pray my dear, quoth my mother, *have you not forgot to wind up the clock?*—God—! Cried my father, making an exclamation, but taking care to moderate his voice at the same time,—*Did ever woman, since the creation of the world, interrupt a man with such a silly question?* Pray, what was your father saying?—Nothing.

---

11. Denis Diderot, *Thoughts on the Interpretation of Nature and Other Philosophical Works*, trans. Lorna Sandler (Manchester: Clinamen Press, 1999), 75–76.

Chapter II

—Then, positively, there is nothing in the question, that I can see, either good or bad.—Then let me tell you, Sir, it was a very unseasonable question at least, —because it scattered and dispersed the animal spirits, whose business it was to have escorted and gone hand-in-hand with the HOMUNCULUS, and conducted him safe to the place destined for his reception.

The Homunculus, Sir, in how-ever low and ludicrous a light he may appear, in this age of levity, to the eye of folly or prejudice,—to the eye of reason in scientifick research, he stands confess'd—a Being guarded and circumscribed with rights:—The minutest philosophers, who, by the bye, have the most enlarged understandings, (their souls being inversely as their enquiries) shew us incontestably, That the Homunculus is created by the same hand,— engender'd in the same course of nature,—endowed with the same loco-motive powers and faculties with us:—That he consists, as we do, of skin, hair, fat, flesh, veins, arteries, ligaments, nerves, cartilages, bones, marrow, brains, glands, genitals, humours, and articulations;—is a Being of as much activity, —and, in all sense of the word, as much and as truly our fellow creature as my Lord Chancellor of England.—He may be benefited, he may be injured,—he may obtain redress;—in a word, he has all the claims and rights of humanity, which Tully, Puffendorff, or the best ethick writers allow to arise out of that state and relation.[12]

Sterne's use of aposiopesis is of this time, and brings the reader into the state of mind of the mother and father, as well as lending agency to all present, including the homunculus.

Sterne's epigraph for the novel is also worthy of an epigraph for this time. Quoting Epictetus: "It is not things that disturb men, but their judgments about things" (1980, 1). The initial three phases of a metaparadigm always address thought in its conceptions as an efficient cause, and the main elements of the formal causal structuring of reality. What we think and how we think shapes the world we live in. The determinism of the fourth phase is not stressed, although it too will have its truths and certainties to enable our human problem-solving.

Even in wholly physical processes that impel evolving life, as was now known in the new understandings of biologists and botanists, i.e., Wolff, Kolreuter and Erasmus Darwin, one could characterize, as did Laurence Sterne, the effects of sound or faulty influences creating individual differences. Caspar Friedrich Wolff sees the evolving of the seed to embryo to infant in animals, and the equivalent process in plants, as the dynamic functioning of coexisting and coordinating parts—vegetative and animal tissue and from them the emergent core organs—fluctuating in energy and their directions of development as they interact with one another. While there is a general model for a species individual, each individual example of a species will differ in the intensity and degree of development as their parts respond "accidentally" to external pressures and with quantitative variability internally to the functioning dynamics of coexistent parts. The conditioning determinism that furthers the general type of a species is not in the forefront as we saw in Montesquieu and will see in Charles Darwin. Rather, with Wolff, one sees

---

12. Laurence Sterne, *Tristram Shandy*, ed. Howard Anderson (New York: W.W. Norton and Company, 1980), 1–2.

"accident" among the interacting parts biologically or botanically, and differences in the quantitative degree of energy in the tissues and emergent organs that emerge from the tissues that generate the distinctive individual specimen. Moreover, we will see an inherent temporality—"growth" as a constant. This will contrast with all fourth phasal conceptual understandings which will stress "duration" of what is studied, with the cessation of the enduring states referenced, but not studied. The biology of Charles Darwin (1809–1882), the geography of Charles Lyell (1797–1875) and Louis Agassiz (1807–1873), and even the shift in ideation by Lamarck (1744–1829), by the fourth phase of this metaparadigm will evidence this shift away from constant "becoming."

Caspar Friedrich Wolff writes in the vein of "becoming" and the individual differences among species:

> Both in plants and animals the particular kinds of development are led by the essential energy of solidification, i.e. coming into presence structurally (Erstarrungsfähigkeit) provided by the nutrient juices. When this occurs with plants or animals there is growth. Where the heart has no activity, and the arteries are not present, and there is no mechanism or other operational principle present, one begins to imagine what will occur as they are manifested. Finally, as these operational mechanisms appear the way of growth is evident. *It is thereby the essential energy of solidification provided by the nutrient juices the is the sufficient principle of every development of plants, as well as animals.*[13]

> When finally in consequence of the spontaneous movements in the body, in the veins, and especially in the folds between the muscles, the fluidic movement is quickened or slowed through compression, as well as in the arteries, by dint of spiritual excitement (geistliche Erregung) or its converse. Since the soul (Seele) is outside of the developing body in its movements of nutrient juices, the soul affects the body with an accessory and accidental developmental principle. Similarly, the air we breathe, which affects the compression degree of the flow of the blood, is regulated internally through an accessory and accidental developmental principle, because the air, as an externality, participates in the processes of bodily development. (Ibid., 62–63 [Par. 248])

The individual differences that emerge in the embryo of a human, and, in the infant, Wolff speculates, can be transferred from the sperm of the father and the contribution of the mother, her egg, which incorporates not only the configurations of body parts, but the spiritual proclivities of both. The idea that the parent's emotional proclivities affect the developing child in the embryo is based, as he claims, by observation. Wolff writes:

> Why the child is similar to the parents can be seen in the sperm of the father and what the mother contributes in this process. From the male sperm, the first developmental parts of the establishment of the body occurs. We know that if the father is choleric or phlegmatic, these emotional traits will affect the flow of blood and nutrients in a similar way in the child. The disposition of the child will be affected by both parents. Both parents are the cause of the child's disposition as the embryo is affected by their spiritual traits. From my observations,

---

13. Caspar Friedrich Wolff, *Theoria Generationis,* Part I (1759) (Leipzig: Wilhelm Engelmann, 1896), 60 [Par. 242]).

I believe the mother's imagination has the greatest influence upon the emergent child's physical development insofar as the nutrient and blood flow intensities in the development of
the embryo. Read my Par. 250[14] in this text for more detail on why both parents affect the
embryo's development. (Ibid., 84–85 [Par. 261])

Wolff, of course, cannot claim to have followed the infant's development of thought and
choice. There is no evidence of actual persons provided. Moreover, *Theoria Generationis*
was his first work at the age of 26, and the embryological insights of Wolff were heightened in future writings, but not in the direction of social-psychological contribution.
That type of evidence will await the third and fourth Modern metaparadigms, and the
spiral building of more intense embryological and mental observation toward this meaningful claim by Wolff. The stress on personal choice and agency in this period of time
must be seen, as with Laurence Sterne, as a bias that influenced the actual observations
of the role of emotions on the physical structures of embryonic development. Speaking
of the mother's imaginative degree of thought and behavior seems as gender-biased addition to what could have been observed by him insofar as actual differences in the child's
individuality from father and mother, and wholly his or her own proclivities of thought
and choice. Nonetheless, as we will see with Freud and his generation, ascribing physical affects to embryonic development by the dispositions and choices of the mother and
father, remains even today as a viable foundation for explaining individuality.

What Wolff does emphasize in the idea of constant development will be studied in
the next initial phase of biology, when we turn to Herbert Spencer. That is, constant
motion, and the force motion generates that will create new combinations of matter. This
dynamical view will not be the focus of Charles Darwin, who is the fourth phase of this
metaparadigm. Rather, the duration of form will be the main inquiry, even while evolutionary change is the core idea. Charles Darwin will argue for the millenniums required
for changes in form that can more successfully adapt to natural conditions. Duration will
be a fourth phase emphasis, while the initial three phases stress quantum changes that are
continual in their preparation and emergence.

---

14. In Par. 250, Wolff writes at greater length about his understanding that one's mental/spiritual
    disposition in thought, and intentional (willkûrlich) living, owes its origins to both parents'
    own thought and intentional life—both emotional and behavioral. See ibid., 66 [Par. 250,
    Anmerkung 2]. Wolff writes there, in part: "Life is determined for the animal by what it thinks,
    and how it intentionally behaves, that is, on account of the activity of the soul. One can
    observe how the consequences of this spiritual activity affect the body as a life process."

Chapter Six

THE SECOND PHASE: DEVELOPING A SYSTEMATIC STRUCTURE FOR GUIDING NEW INQUIRY AND EXPLANATION C.1770–C.1790

It is unfortunate that only after we have spent much time in the collection of materials in somewhat random fashion at the suggestion of an idea lying hidden in our minds, and after we have, indeed, over a long period assembled the materials in a merely technical manner, does it first become possible for us to discern the idea in a clearer light, and to devise a whole architectonically in accordance with the ends of reason. Systems seem to be formed in the manner of lowly organisms, through a *generatio aequivoca* from the mere confluence of assembled concepts, at first imperfect, and only gradually attaining to completeness, although they one and all have had their schema, as the original germ, in the sheer self-development of reason.

Kant, Critique of Pure Reason, 1784 [A 835, B 863]

Kant articulates perfectly the generational change of thought as a new metaparadigm begins—from "the collection of materials in a somewhat random fashion at the suggestion of an idea" in the initial phase to the devise of "a whole architectonically," i.e., systematically. Yet even then the careful inquiry and observation that will solve the problems addressed in the first instance of definitional purview, solidified now into a system, will require the third phase of in-depth inquiry and conflict with traditional modes of generating knowledge.

In this second phase of the second Modern metaparadigm, thinkers in all the arts and sciences were compelled by human precedent in the tasks of new ideation, problem formulation, inquiry design insofar as systematic projection, to take up the new ideation of the first phase. And, how profound this new ideation was! It generated the idea of evolutionary progress, in the species and in each endeavor of the species. Humanity was leveled to equality, and democracy, as we have seen with Rousseau, became a scientifically supported fact. But not to all thinkers. Johann Gottfried Herder, who will be both commended and condemned by Immanuel Kant, his contemporary, wrote a new historiographical model for historical research that argued for monarchical dynasty, a similar argument as presented by Thomas Hobbes—a spiral return as it were, with more evidence.

I will reference Herder both in the philosophy of history and in the new evolutionary science in this chapter. Allow me two epigraphs for the chapter as a whole from

Herder that carry the meaning generally of this second phase of the second Modern metaparadigm:

> Likewise, *was* not that *first, quiet, eternal tree's and patriarch's life* necessary in order to *root* and *ground* humanity in its first inclinations, customs, and institutions?
>
> What were these inclinations? What should they have been? The most natural, strongest, simple ones! The eternal foundation for the education of mankind in all ages [has been]: *wisdom* instead of science, *piety* instead of wisdom, the love of *parents, spouses, children* instead of debauchery, *Life well-ordered, the rule by divine right of a dynasty*—the model for all civil order and its institutions—in all this mankind takes the *simplest*, but also the *most profound delight*. How should all this be developed and passed on, let alone brought forth? in the first place except through the *quiet, eternal power of example* and a range of *[concrete] examples* with their authority about them? By the measure of our own lives, every invention would have been lost a hundred times, springing forth and escaping as in a trance. Who among the immature should have received it? And, who, relapsing too soon into immaturity, should have forced it upon others? Thus, the first bonds of mankind were dissolved at the very outset. The thin, short threads of yore—how could they ever become the strong bonds without which, even after millennia of formation, the human species continues to be dissolved *through sheer weakness?* No! I shudder with joy as I stand before the holy cedar of an original progenitor of the world! Surrounding it are a hundred young, blooming trees already, a beautiful forest of posterity and perpetuation! But look, the old cedar continues to bloom, and its expansive roots are sustaining the whole young forest with their sap and their strength. Wherever the first progenitor may have *gotten* his knowledge, inclinations, and customs *from*, whatever they may be and however paltry—all around him, *a world of present and future generations* has already been formed and fixed in accordance with these inclinations and customs, by nothing more than the *quiet, forceful, eternal contemplation of his divine example!* Two millennia were only two generations.
>
> Johann Gottfried Herder, *Another Philosophy of History* (1774)[1]

As early as 1774 we read in Herder of the continuing religious grounding offered by the Book of Genesis, but here incorporating the idea of generational change, the significance of generational learning and traditional custom. Herder in the first phase of this metaparadigm knew of the work of not only Caspar Friedrich Wolff, but of Laurence Sterne. The epigraph that begins his *Another Philosophy of History* is the same given to Sterne's *Tristram Shandy*, from Epictetus "It is not things but opinions about things that disturb men" (Herder 2004, 3). From Sterne, perhaps, his own power of humanistic analog was reinforced. He saw in Wolff's idea of epigenesis the idea of human self-development based upon successive cultural periods. As early as 1768, in his essay "Über die Reichsgeschichte, Ein historischer Spaziergang" he will coin the term "Zeitgeist," i.e., "spirit of the time."[2] In that concept he will also differentiate between nations, finding

---

1. Johann Gottfried Herder, *Another Philosophy of History and Selected Political Writings*, trans. Ionnis D. Evrigenis and Daniel Pellerin (Indianapolis: Hackett Publishing Company, 2004), 5.
2. Johann Gottfried Herder, *Zur Philosophie der Geschichte, Eine Auswahl in Zwei Bänden* (Berlin: Aufbau Verlag, 1952), 1: 263.

cause in Montesquieu's idea of climate and geography, but, nonetheless focus upon individual choice as he addresses his reader. An excerpt from his 1784 *Outlines of a Philosophy of the History of Man* will be the second epigraph for this second phase of systematic development of ideation, echoing Kant in this recognition of the journey of thought based upon earlier ideas, as he systematized his philosophy:

> An author who produces a book, be it good or bad, in some measure exhibits his own heart to the world, provided this book contain thoughts, which, if he has not invented, and in our days there is little that is new left for invention, he has at least *found,* and made his own, nay, which he has enjoyed for years as the property of his own heart and mind. He not only reveals the subjects that have employed his thoughts or certain periods, the doubts that have occurred to perplex him in his journey through life, and the solutions with which he has removed them, but he reckons upon some minds in unison with his own, be they ever so few, to which these or similar ideas will prove of importance in the labyrinth of life; for what else could excite him to turn author, and disclose what occurs within his own breast to the eyes of a rude multitude?
>
> Herder, Preface, *Reflections on the Philosophy of the History of Mankind*[3]

A. History/Philosophy of History

> When in the Course of human events, it becomes necessary for one people to dissolve the political bands which have connected them with another, and to assume among the powers of the earth, the separate and equal station to which the Laws of Nature and of Nature's God entitle them, a decent respect to the opinions of mankind requires that they should declare the causes which impel them to the separation.
>
> We hold these truths to be self-evident, that all men are created equal, that they are endowed by their Creator with certain unalienable Rights, that among these are Life, Liberty and the pursuit of Happiness—That to secure these rights, Governments are instituted among Men, deriving their just powers from the consent of the governed, —That whenever any Form of Government becomes destructive of these ends, it is the Right of the People to alter or to abolish it, and to institute new Government, laying its foundation on such principles and organizing its powers in such form, as to them shall seem most likely to effect their Safety and Happiness.
>
> The Declaration of Independence, 1776

> We the People of the United States, in Order to form a more perfect Union, establish Justice, insure domestic Tranquility, provide for the common defence, promote the general Welfare, and secure the Blessings of Liberty to ourselves and our Posterity, do ordain and establish this Constitution for the United States of America.
>
> Preamble to the United States Constitution, 1787

The second phase of the second Modern metaparadigm, from the early 1770s through the late 1780s, was a period of extremely principled thinking, where seminal ideas for

---

3. Johann Gottfried Herder, *Reflections on the Philosophy of the History of Mankind* (Chicago: University of Chicago Press, 1968).

reconstructing civil society was central to the thinking of individuals who dwelled upon history, politics, law and other disciplines that had to do with how a society might be better ordered. Systematic models were developed which later became the focus for the difficult work of making them operationally successful.

Immanuel Kant (1724–1804) and Johann Gottfried Herder (1744–1803)

Immanuel Kant's *Idea for a Universal History from a Cosmopolitan Point-of-View* (1784)[4] is as Goethe was later to say of his work "a clean, well-lighted room." Kant offers nine succinct theses that present the systematic structure of how to guide historical action into a developing, progressive experience for the socialization of humankind. Kant's first thesis draws from the evolutionary theory being generated since the late 1750s—"*All natural capacities of a creature are destined to evolve completely to their natural end*" (ibid., 12). From the outset, Kant nests the individual within a species designation, and incorporates the new ideation of progress—here within the older Germanic inclination for individuation theory as taken up by Leibniz from St. Augustine. We will see, just as in Herder's 1784 Preface to *Outlines of a Philosophy of the History of Man*, Kant's appeal to the individual reader. Kant writes under this First Thesis heading, "Observation of both the outward form and inward structure of all animals confirms this of them" (ibid., 12). His second thesis is quite interesting, for he immediately opens up the idea of cultural "generation," and generations of taking up the idea generated by individuals. He writes: "*In man* (as the only rational creature on earth) *those natural capacities which are directed to the use of his reason are to be fully developed only in the race, not in the individual.*" Indicating that thought only occurs in an individual, he conditions this insight with the proviso, as we saw in Herder, that "reason itself does not work instinctively, but requires trial, practice, and instruction in order gradually to progress from one level of insight to another. Therefore a single man would have to live excessively long in order to learn to make full use of all his natural capacities. Since Nature has set only a short period for his life, she needs a perhaps unreckonable series of generations, each of which passes its own enlightenment to its successor in order finally to bring the seeds of enlightenment to that degree of development in our race which is completely suitable to Nature's purpose" (ibid., 13). Whereas Herder saw the Divine as present in custom and tradition, Kant, following Leibniz and individuation theory, sees this Divine purpose in one's ever-present spirit, a metaphysical belief Kant knows cannot be proven. Both he and Herder still remain within Modernism's separate of the Divine from human thought and behavior. The third thesis states this clearly: "*Nature has willed that man should, by himself, produce everything that goes beyond the mechanical ordering of his animal existence, and that he should partake of no other happiness or perfection than that which he himself, independently of instinct, has created by his own reason*" (ibid., 13).

The fourth thesis can be seen as a step toward sociology, indeed a social psychology, a discipline that will begin in this second metaparadigm, c.1820–1865. Kant sees the

---

4.   Immanuel Kant, "Idea for a Universal History from a Cosmopolitan Point-of-View," *Kant, On History*, trans. Lewis White Beck (Indianapolis: Library of the Liberal Arts, 1963), 11–26.

tension between the individual his or her own creative inquiry and the species-demand that it be shared with others, would Nature realize its intentions of progressive development. This fourth thesis is a major step in Western knowledge, offered as one principle in a sequence of systematic principles that this essay intends as a thorough guide for future inquiry and activity. It reads: "*The means employed by Nature to bring about the development of all the capacities of men is their antagonism in society, so far as this is, in the end, the cause of a lawful order among men*" (ibid., 15). Kant explains this thesis: "By 'antagonism' I mean the unsocial sociability of men, i.e. their propensity to enter into society, bound together with a mutual opposition which constantly threatens to break up that society. Man has an inclination to associate with others, because in society he feels himself to be more a man, i.e. the development of his natural capacities.[5] But he also has a strong propensity to isolate himself from others, because he finds in himself at the same time the unsocial characteristic of wishing to have everything go according to his own wish."

As with the other examples of philosophy of history and written history in this phase, one can see a sequence that begins with the dictum "self-realization" by the individual, and leads more and more to a process that demands the individual come to know and build constructively the "society" of that time. Each consequent thesis demands more depth of evidence and understanding. Thus, with Kant's work, the fifth thesis: "The greatest problem for the human race, to the solution of which Nature drives man, is the achievement of a universal civic society which administers laws among men" (ibid., 16). "To do this, each individual must strive to outdo the other insofar as self-realization, as this will benefit the whole of humanity. It is just the same with trees in a forest; each needs the others, since each in seeking to take the air and sunlight from others must strive upward, and thereby each realizes a beautiful straight stature, while those who live in isolated freedom put out branches at random and grow stunted, crooked, and twisted" (ibid., 17).

The sixth thesis raises the idea of the individual more explicitly to the generic limits of his or her natural capacities, in search of exemplary individual who can become leaders toward the civilizing norms to which all must adhere. "Man is an animal which, if it lives among others of its kind, requires a master ... The highest master should be just in himself, and yet a man. This task is therefore the hardest of all; indeed, its complete solution is impossible, for from such crooked wood as man is made of, nothing perfect can be built" (ibid., 17–18). Kant's conclusion is that the laws conceived by generations of men, all individually imperfect, even as they strive for perfection, can be a product of generations

---

5. I depart from Beck's translation only in this sentence from Kant's essay "Der Mensch hat eine Neigung, sich zu vergesellschaften; weil er einen solchen Zustande sich mehr als Mensch, di.i. die Entwickelung seiner Naturanalgen fühlt." See *Idee zu einer allgemeinen Geschichte in weltbürgerlicher Absicht* in *Immanuel Kants Werke, Schriften von 1783–1788*, hrsg. Artur Buchenau and Ernst Cassirer (Berlin: Verlegt bei Bruno Cassirer, 1922), 155. Beck translates "mehr als" as "more than" rather than as "more as." In that Kant introduces the rare expression "vergesellschaften" which means becoming part of a society of others, my translation emphasizes that this in itself is a further realization of his natural capacities—that of language, etc. Karl Marx will pick up this view that "no man is an island" in his similar use of the verbal noun "vergesellschaften."

of thought, and serve to create a civil society for all humankind. He speaks in the seventh and eighth thesis of a world government that is product of all humans over generations; and, that is "nature's secret plan" in the impulsion of human intentions over time. This idealism will be seen in how evolution is conceived by biologists in this period. That is, the human is the apex of evolution, a view challenged in the work of Charles Darwin, Erasmus Darwin's grandson, in the fourth phase of this metaparadigm. Charles Darwin will be the product of a generation that has completely lost faith in the Enlightenment striving for human perfection, or even perfection in the laws. The fourth phase will be compelled to integrate an extremely conservative aristocratic set of principles among the competing systems of governance, monarchy having defeated the emergent republicanism and democracy as the third phase ended. Yet, insofar as metaparadigm theory, collective integration of ideas of the new and the old can vary to a degree. In the case of 1820–1860, aristocracy will have an inordinate weight in those years.

Another of Kant's historical writings in this second phase is *Conjectural Beginnings of Human History* (1786). He structures an architectonic that systematically tracks human origins, progressive socialization with its linguistic and rational competencies, and using Rousseau, reflecting upon the human "social contract" that sought progressively civilized communities. The spirit of this essay is in the initial sentences: "It is surely permissible to insert here and there conjectures into the progression of an historical account, in order to fill gaps in the record. For what precedes the gaps (the remote cause) and what follows them (the effect) give a fairly reliable clue to the discovery of the intermediate causes, which are to make the transition intelligible. But to *originate* an historical account from conjectures alone would seem to be not much better than to draft a novel. Indeed, this could not be called a conjectural history but rather a mere piece of fiction."[6]

Kant raises the reader's awareness with an invitation to use his insights as a guide for further research, he having laid a factual foundation (based upon work done by other scholars, such as Herder, and before him, Rousseau). Kant is quite cognizant of his new platform for "being aware of how and why one is aware." In virtually all his writings, he invites readers to carry his concepts forward with their own inquiry. Moreover, Kant's insertion of his own thought into his books and essays is in the spirit of this new metaparadigm. The first person narrative voice is open about what he knows or does not know—so as to enable the reader to feel his search, and thus as if in a present mutual dialogue, carry the reading into his or her daily thought.

Johann Gottfried Herder does the same as we will see. Herder's 1774 *Another Philosophy of History* offers in its systematic historiography, i.e., what and why history has occurred as it has to date, three sections in sequence. The first section presents the gradual development of the best norms to guide individual education, particularly insofar as being a productive member of a community, and, more broadly, of a society. Herder introduces the idea of progress in how the generic individual should be educated insofar as moral behavior if a peaceful, civic society is to be maintained. As Kant would write, the human

---

6. Immanuel Kant, "Conjectural Beginnings of Human History," *Kant, On History*, trans. Lewis White Beck (Indianapolis: Library of the Liberal Arts, 1963), 53.

spirit is directed toward these ends through his or her choices as each person seeks to realize themselves. The history of cultures who took leadership in the entire history of the human race is introduced. History now is one of shifting leadership, each with a "spirit of the time" as it changed and spread to other cultures. The spirit of this initial section can be seen in these two boldly stated short paragraphs:

> Providence carried along the thread of development—from the *Euphrates, Oxus,* and *Ganges* down *to the Nile* and on toward the *Phoenician coasts*—great strides!
>
> It is seldom without reverence that I leave behind ancient Egypt and the consideration of what it had become *in the history of the human species!* The land where part of the *boyhood* of mankind was formed in its inclinations and knowledge, just as its childhood was in the Orient! The metamorphosis here was as easy and inconspicuous as the genesis had been there. (2004, 11–12)

The second section is a deepening of the narrative of the shifting leadership of world cultures. The initial paragraphs of the second section give the tone of what is to follow:

> The *Roman world-constitution,* too, met its end; and the *greater* the edifice and the *higher* it stood, the greater the plunge when it *fell:* half the world lay in *ruins!* Peoples and continents had lived under the tree,[7] and now, when the voice of the hoy watcher called "Hew it down!" (Compare Daniel. 4:13–14)—how great was the *void!* Like a break in the thread of world events! Nothing less than *a new world* was needed *to heal the tear.*
>
> This [new world] was *the North.* And whatever origins and systems one may devise regarding the condition of *these peoples,* the simplest [account] seems the truest: left alone, they were what one might call *patriarchies such as there could be in the North. Such a climate* made an Oriental *shepherd's life* impossible, since *more burdensome needs bore down* on the human spirit here than where nature provided almost by itself for man, so it was just these *burdensome needs* and the Northern air that *hardened* men more than they could have been hardened in the warm, aromatic greenhouse of the East and the South. Naturally, their condition remained *rougher,* their small societies *more isolated* and *wilder:* but human bonds retained their *strength,* the human *drives* and energy their fullness. Here could arise the land that *Tacitus* describes. And when this Northern sea of peoples became agitated, waves crashed upon waves, peoples upon other peoples! The wall and dam around Rome burst: [the Romans] themselves had pointed out the breaches and had lured [the intruders] to make repairs. Finally, when everything *broke,* how the South was flooded by the North! After all the turmoil and atrocities, what a *new Northern-Southern world!* ...
>
> Not only what *human forces* but what *laws and institutions* did they thus bring onto the *stage of education of the world!* Of course they despised arts and sciences, luxury and refinement—which had wrought havoc on mankind. [Compare Rousseau, *Discourse on the Sciences and the Arts.*] But as they brought *nature* instead of the arts, *healthy Northern intelligence* (*Verstand*) instead of the sciences, *strong* and *good,* albeit *savage customs* instead of refined ones, and as everything *fermented*

---

7. Herder here refers to the metaphor he used in the first section to speak of the will and intentions of the Creator: "Likewise, *was* not that *first, quiet, eternal tree's and patriarch's life* necessary in order to *root* and *ground* humanity in its first inclinations, customs, and institutions? "

together—what a spectacle! How their laws breathed *manly courage, sense of honor, confidence in intelligence (Verstand), honesty,* and *piety!* How their *institution of feudalism* undermined the welter of populous, opulent cities, building up the land, employing hands and human beings, making *healthy* and therefore *happy* people. Their later *ideal, beyond [mere] needs,* tended towards *chastity* and *honor,* [and] ennobled the best part of the human inclinations. Though a *novel,* it was an *exalted novel*: a true *new blossoming* of the human soul. (2004, 32–33)

The rhythm and abstraction of the writing is that of a sequence of principles forming the architectonic whole of which Kant spoke. Herder's praise of feudalism, without any instance of factual example, typifies his political values (in contrast to Kant), and is indicative of the conceptual abstraction one finds in second phase historical writing.

The third section is a synthesis of the first two sections, integrated as a lesson he has presented for himself and the reader. His initial historical scan of periods of Zeitgeist in differing cultures chronologically, and the more in-depth dwelling upon certain of these periods in the second section, such as for example, the arts and sciences of the early Enlightenment under Louis XIV, becomes in the third section a pedagogic for personal improvement. The synthesis being his discussion of his own learning, what his reader could and should learn from these historical presentations, and how this increased knowledge fulfills the Divine plan through individual self-realization. What is distinctive, which we will see in the literature section of this phase, and even in the biological section, is self-confession, a new vision of how to present oneself to the larger world. In the literature section we will touch upon the first autobiographies published since Benvenuto Cellini in a comparable second phase (1563) in the early Modern period. These will be by Rousseau (1770) and Franklin (1770–1790). We read this personalization of Herder, for example, in this paragraph:

The whole (the Zeitgeist) that appears *as a whole* in each of the particulars must be a great one! But in every particular, there is always an *indistinct oneness* that reveals itself and points to the whole. There even *minor* connections can have great *meaning,* and yet entire centuries are mere *syllables*; nations, mere *letters,* and perhaps no more than punctuation marks that mean nothing by themselves but mean *so much* for the easier comprehension of the whole [i.e., God's design of human history—M.E.B.). What are you, *O single human being,* with all your inclinations, abilities, and contributions? In you, perfection is supposed to have *exhausted* all its *aspects?*

The very *limitations* of my earthly point-of-view, the *blindness* of my glances, the *failure* of my ends, the *riddle* of my inclinations and desires, the *defeat* of my powers by the whole that is a *single day,* [let alone] *an entire, year, nation, or age*—precisely this assures me that *I* am nothing, whereas the *whole* is *everything!* ... How miserably *small* it would be if I, *a fly,* could see it all! How little *wisdom* or *great diversity* [would be there] if one who is *stumbling* through the world, who has such trouble holding on to a *single thought,* were never to get *entangled!* [Who am I] on such a stretch, which is nothing, really, but where there are still *thousands of thoughts* and *seeds striving simultaneously;* or in *half a musical bar* of [only] two beats, but where the *heaviest tones* may be winding *towards the sweetest disentanglement*—who am *I* to judge, when I am just *crossing* the great ballroom and eyeing some far corner of the great concealed painting in the dimmest of lights? (2004, 96–97)

When ten years later Herder publishes *Outlines of a Philosophy of the History of Man,* this same symmetry of a chronological overview is present, but one that starts with the evolutionary genesis of life itself upon earth. We read in Herder's initial chapters about the facilitating climate and other conditions for the genesis of life on earth; the evolution of plants, animals and the human, a review of the functions of the parts of animal anatomy as a helpmeet for its development. Chapters I–VII probably convey the meaning of the evolution of all forms of life on earth better than the botanists and biologists insofar as a learning easily accessible to a normal, i.e., unschooled reader. After this systematic architectonic, where nothing is discussed in more depth than required—as a node upon a continuous stem, to be fleshed out in the third phase by other natural historians, biologists and botanists, he turns in Chapters VIII–IX to the human being and their historical communities. He broaches the roles of tradition and custom in meeting their spiritual and animal needs. In Books X–XX we are given the historically chronological architectonic of the national cultures he had sketched in *Another Philosophy of History.* Geography, climate, the story of certain communities who left historical records, important individual leaders are offered, but as cogently as possible, written in a smooth, flowing manner without much detail of times, places or any events. These discussions deepen his earlier argument, but in a way that only provides a better menu for future research—which will await writing in third and fourth phases of this second Modern metaparadigm.

B. Literature

I have resolved on an enterprise which has no precedent, and which, once complete, will have no imitator. My purpose is to display to my kind a portrait in every way true to nature, and the man I shall portray will be myself.

Simply myself. I know my own heart and understand my fellowman. But I am made unlike any one I have ever met; I will even venture to say that I am like no one in the whole world. I may be no better, but at least I am different. Whether Nature did well or ill in breaking the mould in which she formed me, is a question which can only be resolved after the reading of my book.[8]

Jean-Jacques Rousseau, The Confessions (1770)

I should have no Objection to a Repetition of the same Life from its Beginning, only asking the Advantage Authors have of a second Edition to correct some Faults of the first. So would I if I might, besides correct the Faults, change some sinister Accidents & Events of it for others more favourable, but tho'this were deny'd, I should still accept the Offer. However, since such a Repetition is not to be expected, the Thing most like living one's Life over again, seems to be a *Recollection* of that Life; and to make that Recollection as durable as possible, the putting it down in Writing.—Hereby, too, I shall indulge the Inclination so natural in Old Men, to be talking of themselves and their own past Actions, and I shall indulge it, without being troublesome to others who thro' respect to Age might think themselves oblig'd to give me a Hearing,

---

8. Jean-Jacques Rousseau, *The Confessions of Jean-Jacques Rousseau,* trans. J.M. Cohen (London: Penguin Books, 1953), 17.

since this may be read or not as any one pleases. And lastly, (I may well confess it, since my Denial of it will be believ'd by no body) perhaps I shall a good deal gratify my own *Vanity*. Indeed I scarce ever heard or saw the introductory Words, *Without Vanity I may say*, &c. but some vain thing immediately follow'd it. Most people dislike Vanity in others whatever Share they have of it themselves, but I give it fair Quarter wherever I meet with it, being persuaded that it is often productive of Good to the Possessor & to others that are within his Sphere of Action: And therefore in many Cases it would not be quite absurd if a Man were to thank God for his Vanity among the other Comforts of Life—[9]

Benjamin Franklin, *The Autobiography* (1771)

Both Rousseau and Franklin follow a chronological order of chosen events that they deem reveals their character and progress through life, with its gradual public apprecia- tion of who they are. These are stories of how a life should be lived in the new Modern era. The more complex inquiry into the complexities of self and others will follow in the third and fourth phases, but these works are not without the advance in self-knowing that raises "how we are aware of what we are aware of" to new levels.

In both Rousseau's and Franklin's autobiography a self-exposure of possible weakness in character is revealed, yet each of them taking ownership. This is a call to the reader through self-exposure of a like self-examination. This is a public raising of "how each of us is aware of how we are aware"—thus a progression in the norms of self-knowledge. The complex individuality so evidenced is at the threshold of the entry into exploring pre-reflective judgment, the non-conscious consciousness that Franz Brentano and Sigmund Freud will introduce in the third Modern metaparadigm. The autobiography is new on the literary scene of Modern culture, and presents a model that asks readers to review their own lives in its events, to self-correct, or recognize significant actions that should be examples for future behavior in like situations. The example of a novel that is a beacon on both this front, as well as in an even more challenging technique—questioning the objectivity of authorial instruction—will be Goethe's *The Suffering of Young Werther.* Goethe will in this novel create a platform for knowing oneself and one's milieu that is the emergent present in 2020—a critical modernism as we will discuss its still challenging focus upon its discerning the difficulty of determining the truth and proper response to any state-of-affairs, if one is to consider oneself as sufficiently self-aware and effectively responsive.

Johann Wolfgang von Goethe (1749–1832)

Goethe's *The Suffering of Young Werther* (1774) is an individualistic take on the epistolary novels we saw with Richardson and Fielding in the fourth phase of the previous metaparadigm. Goethe, however, does not have dialogues either in a letter that Werther might receive from someone he has corresponded with—he receives none, or in actual dialogue with another with whom he shares a time and place. Rather, the novel is a monologue of the

---

9. Benjamin Franklin "The Autobiography," in *Franklin, Writings* (New York: The Library of America, 1987), 1307–8.

protagonist's individual voice, in letters to a distant friend and descriptions of his live interactions. Moreover, it is the first writing that presents an imperfect narrator, compelling the reader to recognize the misconstructions of objective reality that is presented. The novel is in two parts. There is the discerning narrator of Part I, but then the flawed narrator of Part II. Only the intrusion of the authorial, objective, non-judgmental voice of the impersonal narrator in the last pages of the novel underscore the pathology of Werther that leads to his suicide. Yet, even that suicide indicates another major work, by another author, that is implied to have an unknowingly imperfect author insofar as discerning human motives and how to deal with them and the consequent behaviors these emotions generate.

The first letter to his friend Wilhelm, his distant and never-known to the reader friend, shows us the new narrative technique that had been introduced in the first phase of this metaparadigm, aposiopesis—the staccato, unfinished exclamatory or emotional sentence that draws the reader into the personal struggle of the protagonist. The first paragraph will be the preview of how Werther's actual suffering begins and ends:

> May 4, 1771, How glad I am that got away! Dearest friend, what a thing is the heart of man! To leave you whom I love so much, from whom I was inseparable, and yet be glad! I know you will forgive me. Were not my other attachments deliberately designed by fate to torment a heart like mine? Poor Leonore! And yet I was not to blame. Could I help it that, while her sister's wayward charms provided me with pleasant entertainment, a passion for me grew in her unfortunate heart? And yet, am I wholly without blame? Didn't I encourage her emotions? Didn't I find delight in the wholly sincere expressions of nature which so often made us laugh, however little there was to laugh at? Didn't I, oh, what is man, that he dare reproach himself! I will, my dear friend, I promise you, I will improve; I will no longer, as I have always done, ruminate on the scrap of misfortune that destiny serves up to us; I will enjoy the present, and the past shall be past for me. Of course you are right, my dear friend, there would be less suffering among men if they did not—God knows why they are so constituted—expend so much zeal and imagination in recalling the memory of past ills, instead of enduring an indifferent present.[10]

Werther is unconscionably oblivious of the violation of a sibling relation he has committed with Leonore's sister. This lack of conscience will be the grounds of his final suffering with Charlotte (Lotte) with whom he imagines himself to be mortally in love with in Part II. In Part I, after being attracted to Lotte, he learns she is engaged to Albert—but that doesn't stop his faux passion toward her, his actual flirtatious dalliance. The reader is informed of his distance from Lotte in a few lines. In a letter on July 26, 1771, he includes this sentence:

> Yes, dear Lotte, I will attend to and order everything; just give me more commissions to carry out, and very often. But one thing I beg of you: no more sand on the little notes you write me. Today I swiftly raised your letter to my lips and the sand gritted on my teeth. (1970, 29)

10. Johann Wolfgang von Goethe, *The Sufferings of Young Werther*, trans. Harry Steinhauer (New York: W.W. Norton, 1970), p. 1.

Werther's sense of self is always a balanced stance—not a romantic losing of himself. His flights of romantic engagement, as he wrote in his first letter to Wilhelm, is to enjoy the present as much as possible, regardless of a concern for what may affect others. Lotte, as we will see, feels the same, and walks a careful line so as not to jeopardize her engagement. She, too, is not above leading on the other with faux feelings in general, even as she lets her emotional needs be temporarily satisfied by another, who, in Werther, has a similar relationship to whoever might be that temporary presence. Werther's cold comment is a spontaneous reminder to her and himself of this underlying faux feeling that they share toward each other.

The parting of the two of them that ends Part I is a studied case of civility, indeed, how two flirtations can mutually end with a respect for each other. Werther writes to Wilhelm on September 10 (now four months into the sojourn to Wahlheim) that Lotte as well as Albert met with him on a day that would be their final good-bye. Werther has decided to leave (no reason given) but we can assume that the constant presence of Lotte's fiancé helps him see the fruitlessness of any further association. Lotte, with romantic zeal describes to Werther, in Albert's presence, how at her mother's dying bed, she and Albert were given the mother's blessing. Albert even restrains Lotte to a degree in this scene, saying she is being too emotional. Albert and Lotte part with Werther, she withdrawing her hand from his too tightly grip (ibid., 41–44).

Part II imparts to Wilhelm over the period of another year his efforts to be accepted as a ministerial for aristocratic service, but, also as a ministerial who would be accepted in social engagements with his employers. The sociological frustrations of the Germanies at that time—in the intellectual shadow of both the impending American revolution and the well-read essays of Rousseau reveal Goethe's own democratic leanings. Being a balanced mind and person, Goethe will actually take a position in 1775 with the Duke of Weimar, but until his death, he will have "Auseinandersetzungen" with the Duke over culture, and, most significantly, his initial affair and then marriage to a lower middle-class commoner, Christiane Vulpius.

None of this ability to challenge successfully aristocracy comes into Part II for Werther. Rather, he turns his back on an ability to comply with aristocracy as a ministerial, and fatally chooses to return to Wahlheim (translated as a "choice of home"). He symbolically in Goethe's lesson to the reader puts on the clothes he wore a year before, and now seeks meaning with Lotte, even though she is now married to Albert. Lotte, so flattered, cannot resist teasing the struggle Werther. A sample passage:

> She has been away for a few days to fetch Albert. Today I came to her room; she came toward me and I kissed her hand, overjoyed.
>
> A canary flew from the mirror to her shoulder. "A new friend," she said, coaxing it on to her hand. "I got him for the children. Isn't he adorable? Look at him! When I give him bread he flutters his wings and pecks so nicely. He kisses me too, you see?"
>
> When she held out her lips to the little creature, it pressed against her sweet lips so charmingly, as if it could feel the bliss it was enjoying.
>
> "He shall kiss you too," she said and handed me the bird.—The little beak made its way from her lips to mine, and the peck was like a breath, a foretaste of the delights of love.

"His kiss," I said, "is not entirely free from greed, he is seeking nourishment and returns to you unsatisfied by the empty caress."

"He also takes food from my mouth," she said.—She gave him a few crumbs with her lips, on which the joys of an innocently shared love smiled delightfully.

I averted my face. She should not do this! She should not excite my imagination with these pictures of heavenly innocence and bliss. (Ibid., 61)

Albert evidently is at wit's end to rid himself of the intrusion everyday of Werther's company, and allow himself to give the struggling Werther a pair of pistols for Werther's proposed journey. Albert in giving them to his servant to give to Werther says also "I wish him a happy journey" (ibid., 92). The relief of getting Werther to depart is evident in Albert, and also his giving the quite evident unbalanced Werther at this point the pistols is an understandable exposed motive of hate and vengeance by Albert for the unseemly day-to-day intrusion on his young marriage.

So, Werther will commit suicide with the pistol. And before him by his body one sees, according to the unseen narrator at this point, Gotthold Ephraim Lessing's drama *Emilia Galotti* (1772) open on Werther's desk. Goethe's final tacit criticism of both Werther's decision and Lessing's drama is the conclusion of Lessing's drama, where Emilia has her father take her life in the spirit of how Verginia in ancient Rome allowed her father to take her life, rather than allow her to forcibly become a slave to Appius Claudius, who hungered after her. In Lessing's play, Emilia is in clear and present danger of being seduced by her Prince Appiani. Goethe sees the error in such choices. Lessing actually ends his play by blaming the Prince's minister, Marinelli, for the whole situation, even though Appiani clearly intended to seduce Emilia. Lessing, it seems, refuses to lay blame on the monarchical prince and his governance. Werther, as a character lesson in Goethe's text, should not have allowed himself to succumb to his worst motives. For the reader, this is a misconstruction of what an ethical, seeking, indeed courageous life should be. The book, which introduces the imperfect narrator to literature, was not comprehended in its intentions at Goethe's time. Indeed, it was found at the bedside of a wave of suicides all over Europe.

C. Biology/Botany

Erasmus Darwin (1731–1802)

Darwin's *The Botanic Garden, Part I: the Economy of Vegetation* (1791) had an appreciation as a prelude written in 1788 by W.B. Stephens. Darwin had pursued his inquiries into botany throughout the 1780s, and the poetic form he chose for this 1791 publication came from work done in the 1780s. A more studied scientific inquiry that captured the 1790s third phase inquiry will be looked at in the next chapter, that is his *Zoonomia* (1796–1798).

Erasmus Darwin's *The Botanic Garden, Part I* begins with "An Apology":

It may be proper here to apologize for many of the subsequent conjectures on some articles of natural philosophy, as not being supported by accurate investigation or conclusive experiments. Extravagant theories however in those parts of philosophy, where our knowledge is yet imperfect, are not without their use; as they encourage the execution of laborious experiments, or the investigation of ingenious deductions, to confirm or refute them. And since natural objects are allied to each other by many affinities, every kind of theoretic distribution of them adds to our knowledge by developing some of their analogies.[11]

What he writes could be seen as the most articulate statement for a second phasal awareness of its mode of research, offering a comprehensive set of concepts and things to be inquired into in more depth in the next, third phase, of inquiry. What Darwin says is precisely why an overview that has the theoretical symmetry of a complete knowledge must be established to guide further research.

The poetic text gives this overview of knowing vegetative life. Under each poetic verse is one or more paragraphs that explain the poetic context with facts that are known. Why a poem? Darwin explains that poetry enables analogies, and analogies give bases for the direction of further research:

> The general design of the following sheets is to inlist Imagination under the banner of Science; and to lead her votaries from the looser analogies, which dress out the imagery of poetry, to the stricter, ones which form the ratiocination of philosophy. While their particular design is to induce the ingenious to cultivate the knowledge of Botany, by introducing them to the vestibule of that delightful science, and recommending to their attention the immortal works of the celebrated Swedish Naturalist, LINNEUS.
>
> In the first Poem, or Economy of Vegetation, the physiology of Plants is delivered; and the operation of the Elements, as far as they may be supposed to affect the growth of Vegetables. In the second Poem, or Loves of the Plants, the Sexual System of Linneus is explained, with the remarkable properties of many particular plants. (Ibid., 3–4)

An example of his poetic paragraph and the following explanation is one of his insights into the "Loves of the Plants," or the intentional transmission of germination. Recall, the initial three phases of inquiry in the metaparadigm is intentionality, then followed by determinism. Indeed, the nurture versus nature conflict in the study of organic life separated Charles Darwin who we will study as a fourth phasal thinker, from his grandfather, who reviews that significance of nurture, insofar here as the intentionality of living things who coordinate their lives:

> Come, YE SOFT SYLPHS1 who fan the Paphian groves, And bear on sportive
>   wings the callow Loves;
> Call with sweet whisper, in each gale that blows,
> The slumbering Snow-drop from her long repose;
> Charm the pale Primrose from her clay-cold bed,
> Unveil the bashful Violet's tremulous head;

---

11. Erasmus Darwin, *The Botanic Garden A Poem in Two Parts, Part I: the Economy of Vegetation* (Miami, Fla.: Hard Press, 2016), p. 5.

While from her bud the playful Tulip breaks,
And young Carnations peep with blushing cheeks;
Bid the closed *Petals* from nocturnal cold
The virgin *Style* in silken curtains fold,
Shake into viewless air the morning dews,
And wave in light their iridescent hues;
While from on high the bursting *Anthers* trust
To the mild breezes their prolific dust;
Or bend in rapture o'er the central Fair,
Love out their hour, and leave their lives in the air.
So in his silken sepulcher the Worm,
Warm'd with new life, unfolds his larva-form;
Erewhile aloft in wanton circules moves,
And wose on Hymen-wings his velvet loves.

Darwin then explains:

The vegetable passion of love is agreeably seen in the flower of the parnassia, in which the males alternately approach and recede from the female, and in the flower of nigella, or devil in the bush, in which the tall females bend down to their dwarf husbands. But I was this morning surprised to observe among Sir Brooke Boothby's valuable collection of plants at Ashbourn, the manifest adultery of several females of the plant Collinsonia, who had bent themselves into contact with the males of other flowers of the same plant in their vicinity, neglectful of their own. (Ibid., 174–75)

Chapter Seven

THE THIRD PHASE: MATERIAL INQUIRY INTO THE VERIFIABILITY OF SPECIFIC CONCEPTS, AND CONFLICT OVER THE IMPLICATIONS OF THE FINDINGS C.1790–C.1820

A. Philosophy of History

Immanuel Kant (1724–1804)

Kant's *The Old Question Raised Again: Is the Human Race Constantly Progressing* (1798) is typical of a third phase approach to knowledge. It takes up perspectives that have articulated in the systematic approach—in this case historical development—and probes in depth differing aspects of particular concepts central to the projective understanding that enables problems to be posed, and addressed in action.

Kant states the value of his inquiry:

> As a divinatory historical narrative of things imminent in future time, consequently as a possible representation a priori of events which are supposed to happen then. But how is a history a priori possible? Answer: if the diviner himself creates and contrives the events which he announces in advance. It was all very well for the Jewish prophets to prophesy that sooner or later not simply decadence but complete dissolution awaited their state, for they themselves were the authors of this fate. As national leaders they had loaded their constitution with so much ecclesiastical freight, and civil freight tied to it, that their state became utterly unfit to subsist of itself, and especially unfit to subsist together with neighboring nations. Hence the jeremiads of their priests were naturally bound to be lost upon the winds, because priests obstinately persisted in their design for an untenable constitution created by themselves; and thus they could infallibly foresee the issue.[1]

Kant continues, bringing his emphasis upon perspective and the concepts attached to perspective, up to his present time:

> So far as their influence extends, our politicians do precisely the same thing and are just as lucky as their prophecies. We must, they say, take men as they are, not as pedants ignorant

---

1. Immanuel Kant, "An Old Question Raised Again: Is the Human Race Constantly Progressing?," *Kant, On History*, trans. Lewis White Beck (Indianapolis: Library of the Liberal Arts, 1963), 137–38.

of the world or good-natured visionaries fancy they ought to be. But in place of that "as they are" it would be better to say what they "have made" them—stubborn and inclined to revolt—through unjust constraint, through perfidious plots placed in the hands of government; obviously then, if government allows the reins to relax a little, sad consequences ensue which verify the prophecy of those supposedly sagacious statesmen. (Ibid., 138)

Several new concepts that contribute to the advance in "being aware of how one is aware" are introduced above: (1) a proto-sociology, which one must say for Kant is the awareness of the emergent republican and democratic French revolutionary phases that give examples of how previous societal policies encouraged these new societal movements (ibid., 144); and (2) the specific relationship between a concept, such as progress, regress, revolution, stagnation or some other societal idea as an active state that must be addressed and solved, and, one's view of one's own human nature and one's view of others. This second "be aware of how one is aware" will lead to the social psychology of thinkers such as St. Simon and Karl Marx in the next, fourth phase of this metaparadigm, i.e., 1820–1860.

Kant goes on to examine several other concepts of human nature and its relation to societal structure. "Wickedness" as a basis of who we are cannot sustain us, as it has no problems so articulated that they can be solved. There must be leaders who enable cooperation or humankind will perish through pessimism but such leaders with a populace can be accomplished. "Eudaemonism" where human nature is good, and there is a willingness to try, is seen also as unsustainable, as it depends too much upon individuals who are assumed to be cooperative and helpful. Kant says that "efficient cause" is surpassed because of the uncertainties that do exist in the ebb and flow of human motivation. The "abderitic" point-of-view recognizes the mixed nature of being human, in its evil and good, but to hold this view is to never begin problem-solving because of one's skepticism regarding its possible outcomes. Nietzsche will later speak of this as one being "super-historical." This too has been shown to be false, for there has been "progress" in history. What Kant then goes on to argue is the "progress" can only be seen in "the laws" of humankind. Kant is writing in an era of "constitutionalism," and this does influence his perspective. But, it is humans who make the laws, Kant argues, and so as we have seen in the 1784 essay "Idea for a Universal History from a Cosmopolitan Point-of-View," there must be self-aware and historically aware individuals who can see what laws compel people to adhere to certain behaviors. As in that earlier essay, there are individuals like the Abbe St. Pierre and Jean-Jacques Rousseau who have argued for world government. In this essay Kant is more empirically aware of the difficulties in realizing this because by the time he is writing in 1798, there have already been three phases of the French Revolution—constitutional monarchy, democracy and republicanism. He writes cautiously:

The answer is not by the movement of things *from bottom to top,* but *from top to bottom.* To expect not simply to train good citizens but good men who can improve and take care of themselves; to expect that this will eventually happen by means of education of youth in the home, then in the schools on both the lowest and highest level, in intellectual and moral culture fortified by religious doctrine that is desirable, but its success is hardly to be hoped for. For while the people feel the costs for education of their youth ought to be borne, not by them, but by the state, the state for its part has no money left for the salaries of its teachers who are capable and zealously devoted to their spheres of duty, since it uses all its money for war. Rather, the whole mechanism of this education has no coherence if it is not designed in agreement with a well-weighed plan of the sovereign power, put into play according to the purpose of this plan, and steadily maintained therein. To this end it might well behoove the state likewise to reform itself from time to time and, attempting evolution instead of revolution, progress perpetually toward the better.

Nevertheless, since they are also human beings who must effect this education, consequently such beings who themselves have been trained for that purpose, then, considering this infirmity of human nature as subject to the contingency of events which favor such an effect, the hope for its progress is to be expected only on the condition of a wisdom from above (which bears the name of Providence if it is invisible to us) but for that which can be expected and exacted from *men* in this area toward advancement of this aim, we can anticipate only a negative wisdom, namely, that they will see themselves compelled to render the greatest obstacle to morality—that is to say war, which constantly retards this advancement—firstly by degrees more humane and then rarer, and finally to renounce offensive war altogether, in order to enter upon a constitution which by its nature and without loss of power is founded on genuine principles of right and which can persistently progress toward the better. (Ibid., 152–53)

Kant sees rational law as gradually developed so that gradually those in a society of these laws as they are improved toward humane and effective cooperation can guide the problem-formulations and problem solutions that enable "progress." Kant speaks as one who in a third phase of a metaparadigm is at work in problem solving, attempting to see what needs to be done. In his Europe these efforts are being attempted by all sovereign powers in the 1790s, be it France, England, the Germanies or the lesser nations. Throughout the West, including the United States, such efforts of careful identification of spheres of needed improvement and the address of those spheres is taking place. And what of the individual education of these individuals throughout the West? Literature will address these concerns in the depth, and conflict among perspectives for solution, that is the third phase.

B. Literature

Johann Wolfgang von Goethe (1749–1831)

When I read aloud to you, is it not as though I were speaking directly to you? The written and printed words take place of my own thoughts, my own heart. Would I make the effort to talk if a little window had been set into my forehead or breast, so that the person to whom I communicate my thoughts and emotions, one by one, would know all the while, in advance, what I was driving at? If someone glances into the book I am reading, I always feel as if I were being pulled apart.

Goethe, *Elective Affinities* (1809)[2]

---

2. Johann Wolfgang von Goethe, *Elective Affinities* in *The Sufferings of Young Werther and Elective Affinities*, trans. Elizabeth Mayer and Louise Bogan (New York: Continuum Publishing Company, 1991), 155.

This is spoken by Eduard to his wife Charlotte, two of the main protagonists in this novel. The allusion is actually taken from a classical myth of the arch-critic Momus, who criticizes the craftsman-god Hephaestus for how man was created.[3] Momus represents a God, one who originated in Asia Minor, who was not only the most schooled on aesthetics, but motives that were behind behavior. This observation by Eduard, taken directly from the Momus fable in Babrius, is evidence of how "being aware of how one is aware" was a reflective possibility in the ancient world. As I indicated earlier in this text, careful reading of Aeschylus, Sophocles, Aristophanes and Euripides shows the ability of the personae to see themselves as of two minds in situations, as well as to see the intentions of others, even when these individuals put forward a guise. The return of this public affirmation of such human reflective/perceptive thought marked the mixed motives of power and secular success of Modernism, and was more on the forefront of how literature addressed the everyday of its characters.

Goethe wants to have Eduard coach his wife into looking more deeply for states-of-mind in himself, and thus also recognizes that whenever people speak or read, it is their personal understanding of what they read. This recognition by Goethe enables him to see his own behavior over time, and his states-of-mind in the moment with greater clarity than the non-reflective person. Goethe's own autobiography is written in the period soon after writing the *Elective Affinities,* probing only several decades of his life—mainly those of is coming into "formal reasoning" (ages 9 through 14, according to Rousseau and Piaget), and into his University years. This is when he began to monitor his own responses to people and situations in this ability to see his own mixed motives. The personal encounters he documents all have indications in the narrative of his own, often ironic, personal interactions with or observations of others.

Eduard throughout the *Elective Affinities* draws further away from Charlotte, and with the entry of the younger Ottilie, finds a soulmate. Charlotte, in turn, lives in this tepid relation to Eduard in a mutual manner, she finding a pure love in the entry of the Captain to their home. The novel is one of the struggle to find a domestic solution, while remaining aware of their separate deeper loves. This novel, one can say, picks up where *The Sufferings of Werther* (1774) entered Part II, but the self-insight, and insights into others by Eduard and Charlotte create a sane, relatively successful path toward their future lives. The problem of Werther is made indirectly more explicit, and Goethe may have wished readers would return to see the errors of Werther's loss of self-insight more clearly. Eduard and Charlotte are each in their second marriage, one that began with mutual love. Eduard and Charlotte, out of mutual respect and honoring their initial love, remain married until his death. The Captain is renounced by Charlotte, who sees that the moral decorum of her marriage whose chrysalis was love, should not be revoked. Eduard remains an obliging spouse, even as his love for Ottolie, who remains in their home, is explicit every day. Toward the end of the novel, Charlotte speaks of the lesson

---

3. See *Babrius and Phaedrus,* ed. Ben Edwin Perry (London: William Heinemann, 1965), 75 [Fable 59]. See also Lucian, "Hermotimus, or the Rival Philosophies," *The Works of Lucian of Samasota,* trans. H. Fowler and F.G. Fowler, 4 vols. (Oxford: Clarendon, 1905), 2: 52.

Goethe teaches us in the novel, a lesson of constant insight into "being aware of how we and others are aware, in all its complex moments." Charlotte says:

> "While life is carrying us along with it," she said, "we imagine that we act from our own motives and choose what we do and what we enjoy; but, if we look more closely we all find that we are actually compelled to carry out the ideas and tendencies of our time."

> "That is true," the Tutor replied. "Who, after all, can resist the force of the tendencies of his period? Time moves on, and with it opinions, ideas, prejudices and fashions. If the early years of a young man fall in a period of transition, we can be sure that he will have nothing in common with his father. If the father lived in a period which tended toward acquiring a good deal, toward protecting this property, restricting and confining it, and in the sequestration from the world securing its full enjoyment—the son will be inclined to expand, to communicate, to extend and to open up what was closed."

> "Whole periods resemble this father and son you have described," Charlotte agreed. (Ibid., 279–80)

Goethe sees not only periods of time, the "Zeitgeisten" of which his contemporary Herder wrote, but the "transition times" where a mixture of attitudes co-exist. Charlotte has chosen fidelity to a husband where love has cooled, despite her depth of love for the Captain. He will marry, for position and wealth. This is the emergent conservatism of the first decade of the nineteenth century for ministerials, and the class that will be called bourgeois, which included many of them. Aristocrats in the Germanies, particularly the Duke of Weimar's mother, and to a great degree, he himself, allowed ministerials to intermix socially with their own in gatherings.[4] The *Elective Affinities*, as Charlotte states, makes evident that social pressures are the final word in milieus, and the sane, balanced person adapts. Here is, again, the proto-sociology and its social psychology.

New disciplines begin to emerge between 1790 and 1820 on the strength of the careful, in-depth scholarship in these years. Politically, many commoners were ennobled given their prowess in improving the structure of governance. This was particularly true in the Prussian approach toward their military, Frederick III being quite open to talent within his domain. Between 1807 and 1819 August Neidhardt von Gneisenau and Gerhard Scharnhorst, born commoners, generated permanent conscription laws and a national guard. Hundreds of thousand soldiers were thus guaranteed. But, also officers from the common people (university trained upper middle class mostly).

C. Biology/Botany

Erasmus Darwin (1731–1802)

> The late Mr. David Hume, in his posthumous works, places the power of generation much above those of our boasted reason; and adds, that reason can only make a machine, as a

---

4. See Friedrich Sengle, *Das Genie und sein Fürst, Die Geschichte der Lebensgemeinschaft Goethes mit dem Herzog Carl August* (Stuttgart: J.B. Metzlershe Verlagsbuchhandlung und Carl Ernst Poeschel Verlag, 1993), 18–19.

clock or ship, but the power of generation makes the maker of the machine; and, probably from having observed, that the greatest part of the earth has been formed out of organic recrements; as the immense beds of limestone, ironstone, coals, from decomposed vegetables; all which have been first produced by generation, or by secretions of organic life; he concludes that the world itself might have been generated, rather than created; that is, it might have been gradually produced from very small beginnings, increasing by the activity of its inherent principles, rather than by a sudden evolution of the whole by the Almighty fiat.—What a magnificent idea of the infinite power of the Great Architect! The Cause of Causes! Parent of Parents! Ens Entium!

For if we may compare infinities, it would seem to require a greater infinity of power to cause the causes of effects, than to cause the effects themselves. This idea is analogous to the improving excellence observable in every part of the creation; such as in the progressive increase of the solid or habitable parts of the earth from water; and in the progressive increase of the wisdom and happiness of its inhabitants; and is consonant to the idea of our present situation being a state of probation, which by our exertions we may improve, and are consequently responsible for our actions.

The efficient cause of the various colours of the eggs of birds, and of the hair and feathers of animals, is a subject so curious, that I shall beg to introduce it in this place. The colours of many animals seem adapted to their purposes of concealing themselves either to avoid danger, or to spring upon their prey. Thus the snake, and wild cat, and leopard, are so coloured as to resemble dark leaves and their lighter interstices; birds resemble the colour of the brown ground, or the green hedges, which they frequent; and moths and butterflies are coloured like the flowers which they rob of their honey.[5]

Darwin, writing this in Volume One of his *Zoonomia* (1794) states clearly the increase in our knowing, that is "being aware of how the immediate causes of persons, places, and things have causes themselves that have heretofore not been known." He emphasizes in the above second paragraph that "the progressive increase of the wisdom and happiness of its inhabitants ... is consonant to the idea of our present situation being a state of probation, which by our exertions we may improve, and are consequently responsible for our actions." His intent in these words is to say that the Divine meant us to discover and employ to our advantage the formal causal sequences that enables specific cause and effects to be generated. This emphasis upon our personal agency, especially in the third phase of this metaparadigm speaks to the thorough inquiry into botanic and biological issues by researchers, moreover, to "improve" which means to solve problems. Among the problems of course is to effectuate an enhanced hybridization and husbandry. Charles Darwin, his grandson, will cast doubt upon the long-range effects of planned hybridization or husbandry. I will come to that in the next chapter, where personal agency in interjecting one's ideas into cultivation and breeding is discouraged by Charles Darwin, and natural determinism encouraged.

---

5. Erasmus Darwin, *Zoonomia; or, The Laws of Organic Life in Three Parts,* Complete in Two Volumes (Boston: D. Carlisle, 1803), I: 400–401.

The natural effort of plants and animals to improve their persistence in the competition for survival is recognized by Erasmus Darwin in the above third paragraph. This
genetic change shows an "intentional"[6] change in the species which is also discussed as
either "soft evolution" or "epigenetic inheritance," which Erasmus Darwin's contemporary, Jean-Baptiste Lamarck (1744–1829), believed would be inherited in descendants.

Epigenetic inheritance has been argued by scientists including Eva Jablonka and Marion
J. Lamb to be Lamarckian.[113] Epigenetics is based on hereditary elements other than genes
that pass into the germ cells. These include methylation patterns in DNA and chromatin
marks on histone proteins, both involved in gene regulation. These marks are responsive to
environmental stimuli, differentially affect gene expression, and are adaptive, with phenotypic
effects that persist for some generations. The mechanism may also enable the inheritance of
behavioral traits, for example in chickens, rats and human populations that have experienced
starvation, DNA methylation resulting in altered gene function in both the starved population
and their offspring. Methylation similarly mediates epigenetic inheritance in plants such as
rice. Small RNA molecules, too, may mediate inherited resistance to infection. Handel and
Romagopalan commented that "epigenetics allows the peaceful co-existence of Darwinian
and Lamarckian evolution."

Joseph Springer and Dennis Holley commented in 2013 that Lamarck and his ideas were
ridiculed and discredited. In a strange twist of fate, Lamarck may have the last laugh.
Epigenetics, an emerging field of genetics, has shown that Lamarck may have been at least
partially correct all along. It seems that reversible and heritable changes can occur without a
change in DNA sequence (genotype) and that such changes may be induced spontaneously
or in response to environmental factors—Lamarck's "acquired traits." Determining which
observed phenotypes are genetically inherited and which are environmentally induced remains an important and ongoing part of the study of genetics, developmental biology, and
medicine.[7]

The ascription of "intentionality" to plants, and the possible inheritance of their learned
abilities, is a third phasal inquiry in our contemporary metaparadigm, i.e., "now." That
this third phasal inquiry by Lamarck and others which took place in the last decade
of the eighteenth century has seen a "spiral return" today is not surprising insofar as
the theses I am pursuing. The apparent "sessile" immobility of plants, yet their active
agency in exploring better survival, is discussed in the article by Michael Marder. A few
paragraphs suffice to show the perspective of Marder:

One of the most obvious features of plant life is the fact that plants are sessile. All too often,
sessility has been mistaken for the plants' immobility and impassiveness, with the notable
exceptions of rapid movements observed in *Mimosa pudica* or *Dionaea muscipula*. This is a quin-
tessentially modern prejudice, resulting from the exclusive identification of movement with
locomotion. Aristotle, to his credit, recognized that the latter is only one of four types of

---

6. https://www.ncbi.nlm.nih.gov/pmc/articles/PMC3548850/
7. https://en.wikipedia.org/wiki/Lamarckism#Transgenerational_epigenetic_inheritance

movement, the other three being growth, decay and change of state (metamorphosis), all of which are present in plant life. It is evident that the fixedness of plants is an impression-istic mistake, given their lateral and vertical extensions both above and below ground level. Although they appear to be anchored in a place, plants incessantly explore their environ-ments, maximizing their exposure to sunlight, avoiding or growing toward the roots of their neighbors and monitoring and responding to changing environmental conditions. (Marder 2012, 1368)

Marder employs the phenomenological method in his substantiation of "intentional" movement:

The chief cause behind the illusion of plant immobility is the difference in the time scales of human and plant lives. In everyday settings it is impossible to perceive the growth of plants, since many plant responses may take days or even weeks. From the phenomenological van-tage point, not only the sense of place but also that of time is indexed to the subject who experiences it. In contrast to the objective "clock time," Husserl, who stood at the origins of this intellectual movement, put emphasis on "internal time consciousness," or on how subjects experience the passage of time as either fast or slow depending, for example, on their mood at any given moment. It is likely that such variations in temporal perspectives are not only inter-personal but also extend to cross-species and cross-kingdoms differences. If the phenomenology of plant intelligence is relative to the capacities of plants, then these, too, must be relative to the specific temporal framework, wherein these capacities are enacted.

One area where there is a partial overlap between the internal time consciousness of an-imals and plants is time estimation with the help of circadian clocks. Besides the fact that the same molecular mechanisms permit plants and animals to exploit circadian clocks, leaves of some plants, such as *Lavatera cretica*, can anticipate the direction of sunrise, even after they have been prevented from solar tracking for several days. The combination of memory and anticipation is consistent with the phenomenological description of time as the retention of a past "now-moment" and the projection into a future "now-moment" by a conscious subject. The sense of place remains incomplete without this, its experiential temporal dimension.

Plant intelligence entails, at the most basic level, the subjective constitution of lived space and time by the plants themselves. Plant behavior is marked by a successful (from the practical or pragmatic point of view) orientation in local environment, taking into account minute changes in temperature, humidity gradients and so forth. One of the reasons behind this suc-cess is that plants grow not so much in opposition as in contiguity with the ecological niche they inhabit, as evidenced by the maximization of their surface exposure. A rooted mode of being and thinking is, then, characterized by extreme attention to the place and context of growth and, hence, by a sensitivity that at times exceeds that of animals. (Ibid., 1369–70)

Marder's concluding paragraph is his argument that what has been discerned in some depth establishes a "phytophenomenology" to be pursued by others for future discov-eries. The article implies "descendant" acquisitions of intentional abilities in plants. This would have to be furthered with the study of plants being transplanted to differing environments. Nonetheless, Marder presents a justification that harks back to Erasmus Darwin's approach, as we will study further:

This article provided no more than the prolegomena to the fruitful interdisciplinary combination of phenomenology, botany and population ecology—an approach we may term
phytophenomenology. Each of the phenomenological themes under discussion here merits
further consideration, so as to inform current debates surrounding plant decisions, choices and
behaviors that are not determined in linear and mechanistic ways. A crucial methodological
advantage of phytophenomenology is that it neither treats plants as passive objects (or quasi-
mechanical structures relegated to the background of animal life) nor accepts the Western
metaphysical equation of subjectivity with autonomy, unity, individuality, personhood or will.
The centerpiece of this approach is the phenomenological concept of intentionality and its
relevance to plant life: the directedness-toward of intentionality as a general descriptor of
behaviors characteristic of sessile and mobile beings; the dispersion of plant intentionality
in vegetal sentience and bio-attention; the spatial and temporal construction of plant world
through a network of dispersed intentionalities; the appropriateness of modular development
to this dispersion; and, finally, plant co-intentionality and clashing intentionalities as theoretical descriptions of communication, kin recognition and cross-species / cross-kingdoms
interactions. A supplement to the cognitive (information-processing), evolutionary and ecological perspectives on plant intelligence, phytophenomenology is thus capable of synthe-
sizing large amounts of scientific data into a coherent explanatory framework. (Ibid., 1372)

Erasmus Darwin's *Zoonomia* begins with a study of motion and its causation in the
plant and animal world. This dynamical topic is understood by Darwin as the energetic
core of animate life, therefore, the best beginning. His language is that of discerning
"intentionality" that is, "a directed act of consciousness." Darwin doesn't use that term,
an understanding of being conscious that is phenomenological, the new philosophy of
the late nineteenth century—a first phasal conceptual formulation by Franz Brentano.[8]
Phenomenological intentionality is a willed direction of attention, but not the overlay,
necessarily of a "reflective choice." Seeing in plants and animals a constant movement
of their animate abilities, initially non-reflective (Brentano 1973, 153–54), toward that
which can foster survival is the level of what is understood as intentionality by phenomenological inquiry. Darwin uses the term "spirit' and its "volition." He stresses that plants
and animals have "spirit," and that is the central efficient cause of its volitional movement and their effects:

The whole of nature may be supposed to consist of two essences or substances; one of which
may be termed spirit, and the other matter. The former of these possesses the power to communicate or produce motion, and the latter to receive or communicate it. So that motion,
considered as a cause, immediately precedes every effect; and considered as an effect, it
immediately succeeds every cause. And the laws of motion therefore are the laws of nature.

---

8. Franz Brentano, *Psychology from an Empirical Standpoint*, trans. Antos C. Rancurello, D.B. Terrell,
    and Linda L. McAlister, ed. Oskar Kraus and Linda L. McAlister (New York: Humanities
    Press, 1973), 88–89, 180–81, especially notes on those pages. See also, Robert Sokolowski on
    Husserl's deepening of the concept of intentionality in the third phase of this metaparadigm,
    Introduction to Phenomenology (Cambridge: Cambridge University Press, 2000), 8, 12, 20–21, 40,
    147–48 and 216–17.

The motions of matter may be divided into two kinds, primary and secondary. The secondary motions are those, which are given to or received from other matter in motion. Their laws have been successfully investigated by philosophers in their treatises on mechanic powers. ... The primary motions of matter may be divided into three classes, those belonging to gravitation, to chemistry, and to life; and each class has its peculiar laws. (Darwin 1803, I: 1)

Darwin goes on with the whole of Volumes One and Two, over 900 pages in differing categories and sub-categories of motion and its effects on plants and animals. Speaking of "volition" in the sense of what will be understood in the spiral advance of the next third phase of Husserlian "intentionality" he writes:

A ... difficulty may have arisen from the confined use of the words "to will," which in common discourse generally mean to choose after deliberation; and hence our will or volition is supposed to be always in our own power. But the will or voluntary power, acts always from motive, as explained in Volume One, Section XXXIV and in Volume Two, Classes III and IV, which motive can frequently be examined previous to action, and balanced against opposite motives, which is called deliberation. At other times the motive is so powerful as immediately to excite the sensorial power of volition into action, without a previous balancing of opposite motives, or counter volitions. The former of these volitions is exercised in the common purposes of life, and the latter in the exertions of epilepsy and insanity. (Darwin 1803, II: 277)

Darwin observed multiple cases of the volition in humans he understood as before reflective choice. This in-depth inquiry and observation is more than definitional or to create a symmetrical outline for study, as in the initial two phases. He writes of the "diseases of the spirit," whose movement of mind and body are prior to reflective choice. One of the diseases of the spirit, of many he observes and details, is "sentimental love":

The passion of love produces reverie in is first state, which exertion alleviates the pain of it, and by the assistenace of hope converts it into pleasure. Then the lover feels solitude, less this agreeable reveries should be interrupted by external stimuli, as described by Vergil.

Tantum inter densas, umbrosa eacumina, sagos

Assidue veniebat, ibi haec incondite solus

Montibus et sylvis sudio jactabat inani.

When the pain of love is so great, as not to be renewed by the exertions of reverie, as above described, as when it is misplaced on an object of which the lover cannot possess himself, it may still be counteracted or conquered by the stoic philosophy, which strips all things of their ornaments, and inculcates "nil admirari," of which lessons may be found in the meditations of Marcus Antoninus. The maniacal idea is said in some lovers to have been weakened by the action of other energetic ideas; such as have been occasioned by the death of his favorite child, or by the burning of his house, or by his being shipwrecked. In those cases the violence of the new idea for a while expends so much sensorial power as to prevent the exertion of the maniacal one; and new catenations succeed. On this theory the lover's leap, so celebrated by the poets, might effect a cure, if the patient escaped with life.

The third stage of the disease I suppose is irremediable; when a lover has previously been much encouraged, and at length meets with neglect or disdain; the maniacal idea is so painful as not to be for a moment relievable by the exertion of reverie, but is instantly followed by furious or melancholy insanity; and was lately exemplified in Mr. Hackman, who shot Miss Ray in the lobby of the playhouse. So the poet describes the passion of Dido,

Moriamur inultae?—

At moriamur, ait,--sie, sie, juvat ire sub umbras!

The story of Medaea seems to have been contrived by Ovid, who was a good judge of the subject, to represent the savage madness occasioned by ill-requited love. Thus the poet,

Earth has no rage like love to hatred turn'd
No Hell a fury like a woman scorn'd.
                    Dryden

Hence it appears that though sentimental love does not so frequently arise spontaneously in female bosoms, yet that it is liable to become as violent, when it has been excited by the courtship of the other sex, and though, when it is rejected, after courtship has produced it, it is not always succeeded by such violent effects as those above mentioned; which may be ascribed to the greater modesty and reserve of their education, yet the disappointed passion is liable to prey upon their minds even to the hazard of their lives, of which I have witnessed two instances, in both which the effects approached to that occasioned by great grief. See Moerer, Class III. L.2. 10.[9]

One of these ladies, about 30 years of age, was deserted by an Irish gentleman, who was soon to have married her. She was suddenly seized with a stupor, which by those who were not acquainted with the cause, was mistaken for a kind of apoplexy; she gradually recovered as to apply to her usual habits of life and in four or five years regained her cheerfulness, and married another man. The other was affected with long stupor, loss of digestion, and total inability of mind and body, which continued a year or two, and from which she also gradually recovered. (Ibid., II: 312–14)

Darwin offers us a phenomenological account of the movement of the mind and its interaction with emotion as well as the physical body. This is not an excess and loss of focus by Darwin, nor poetic license—as even his poetic accounts of botany in the previous phase are annotated with conceptual acuity. Quoting ancient sources of poets is rather a brilliant epigraph for his own inquiry into the immediacy of volitional states that are pre-reflective, affecting ideation and movement, as well the movement of ideas. This comprehensiveness of inquiry will return in a spiral manner, that is more depth a detail, to the third phase each century in its new metaparadigm. We will see this same

---

9. Darwin directs us here to his study of "grief" in its effects; ibid., 317–18.

understanding of pre-reflective volition in the psychology of Franz Brentano and his student, Sigmund Freud. Freud's studies enter into this same examination of both female and male "hysteria" in the first decade of the twentieth century.[10] Freud's "diseases of the soul" are of the same content and explanation as the seminal inquiries of Erasmus Darwin a century before.

---

10. See especially, "Hysterical Phantasies and Their Relation to Bisexuality," in *Sigmund Freud, Collected Papers,* trans. Joan Riviere, 5 Volumes (New York: Basic Books, 1959), 2: 51–58.

Chapter Eight

THE FOURTH PHASE: INTEGRATING THE NEW FOUR CAUSAL UNDERSTANDINGS WITH THE TRADITIONAL C.1820–C.1860

The traditional four causes that preceded Modernism were still alive in the minds and actions of a majority of the Western population. The final cause, formal cause, the efficient causes that constitute the patterns pursued as thorough research that realizes the formal cause and the material cause in each discipline of this period of time sought to integrate with the new perspectives that had begun in the 1760s. That meant traditional monarchy as an absolute authority, or constitutional monarchy at best, the Divine, even as a "proximate" presence, and the traditional self-limitations on the nature of inquiry into material reality, sought to integrate all into its perspective, and in so doing had to find a common ground with those who pursed the new four causes. The new four causes were the final cause of the human in and for themselves, the formal cause of a systematic inquiry into every facet of the material realities within which humans had always and now lived, and the efficient causes in all their complexity of coexisting, mutually affecting, organic and inorganic life.

There had been two major revolutions in Europe in the initial three phases of this metaparadigm between the 1760s and 1820. Both republicanism and democratic governance had emerged as national systems. Through the third phase pragmatism of inquiry and agentive action what had been theory became accustomed possible practice. Moreover, the traditional position in governance smarted under the successful rebellions. With their successful defeat of Napoleon in France in 1815, and the chastening of the United States by England with the War of 1812, monarchs were more conservative than before, much less willing to integrate new thought through the change of laws. The Congress of Vienna where the leaders of the major European nations met between 1814 and 1815, coordinating their conservative policies until 1820, agreed, except for England, to a Holy Alliance of orthodox religion as national religion in their respective countries in 1815. They agreed in 1820 to the Troppau Protocol, where their armies would put down republican or democratic rebellions in any monarchy in Europe that requested aid. England eschewed both the Holy Alliance and the Troppau Protocol, yet the Regency period with the future George IV as Regent for George III, was a Tory regime until 1822. In 1819 England's government passed the Six Acts of national censorship in response to a worker's strike. In the same year the Germanies passed the Carlsbad Decrees that led to police surveillance and censorship. How then did the search for mutual understanding between traditional views and the new consciousness evolve between 1820 and 1860?

The First Industrial Revolution had produced a large steam engine in 1760 (James Watt) that powered factories 24 hours a day. The savings by manufacturers enabled them to undercut artisan labor, and both the English and French governments gave privileged attention to these upper-middle-class businessmen. With the invention of the steam-powered locomotive in 1830, England and then France took leadership in the new industries and means of transportation. Prussia started a customs union in 1837 to facilitate both railroad technology and the new industrial factories that comprehended 26 of the 39 principalities of the German Confederation. Thus by the republican and democratic revolutions of 1848, three of the five major nations of Europe supported new technological sciences that while supporting business, also were supported by governments in the way of funding, and thus reinforcement of new knowledge in the scientific disciplines. The revolutionary years of 1848 through 1849 saw aristocrats back in charge, with only slight concessions to the manufactures of the mature Industrial Revolution. But, the concessions gave this "upper bourgeois" class of individuals legislative power in the Germanies, France and in England. We will see between 1850 and the early 1860s more integration between the traditional and the new, where national policies began to be dictated by commoners. This enhanced cooperation will lead to the Second Industrial Revolution in the 1860s.

This partnership with the manufacturing class and aristocrats can be called Neo-Absolutism. Between 1850 and the late 1860s it was coupled with funding for the rebuilding of capital cities, each nation seeking to show the new public expenditures, buildings, gardens and other points of pride. This led to what could be called Neo-Imperialism, where the populace of a nation supported a news drive for colonies around the globe. In 1850, 55 percent of the globe had been colonized by European nations. Between 1850 and 1914, 85 percent of the world's non-European populations became colonies of the European nations. The partnership of manufacturers and aristocrats generated a foreign policy, as well as domestic public policy—taxes, military, commerce—dominated by the perspectives of the manufacturing class. This, of course, led to World War I.

Socialist thought and its expression in political parties began in 1820, and by the early 1860s was a third force in Europe sharing emergent power with the capitalist manufacturers, and, with the growing weakness after the 1860s, of monarchy.

The need and outreach for integrating with the increasingly coherent principles of socialism, capitalist big business and monarchy from 1820 through the 1860s was accomplished through skilled political agents of the differing political visions. Frederick III and his son Frederick IV, Kings of Prussia, facilitated the First Industrial Revolution there, and with the economic advising of Friedrich List, and the productive growth of the Customs Union of 26 northwest and north central states, created a lasting, successful partnership with manufacturers. Otto von Bismarck carried this forwards successfully from the 1860s onwards. In England, the Whig party was the party of big business, and powerful political figures as Lord Palmerston generated a worldwide development of English dominance in the First Industrial Revolution. In France, Louis Philippe, the Citizen King between 1830 and 1848 made France second in the world in First Industrial Revolution products, as well as in railroads. Even in Russia, Nicholas I, between 1825 and 1855, followed by Alexander II, brought cooperative manufacturing relationships with France in particular, but also Prussia. Only Austria lagged behind, and that was to blame on one of the foremost political figures of the fourth phase—Klemens von Metternich. But, his fall in 1848 led to attempt to form bonds with their manufacturing class—with railroad building and industrial manufacturing.

On the socialist side, France inherited the legacy of social-democratic governance from the period of Robespierre, which the urban populations of France never forgot. Socialist theory was brilliantly articulated by St. Simon in 1820, and was continued theoretically by Charles Fourier in the next decades. The socialist Louis Blanc was perhaps the leading political figure in the early French Revolution of 1848. The Social Democratic approach to democracy is heavy on government funding, and, Napoleon III, who kept France growing industrially in the Second Industrial Revolution until he abdicated in 1870, established two national banks, which served industrialist republicans as well as cultivating a wholly democratic National Assembly.

So what we will see in this fourth phase of the second Modern metaparadigm from the 1750s through the early 1860s is the sharpening of the sovereignty ideas, yet the ongoing attempts to cooperate across political ideologies.

A. Philosophy of History/History

Leopold van Ranke (1795–1886)

Among the major historians in the Germanies at this time was Leopold von Ranke (1795–1886) whose political values were supportive of monarchy. Georg Wilhelm Friedrich Hegel (1770–1831), whose political values were of the ministerials, those commoners who were elevated, some to lower aristocracy, because of their University training, and subsequent success in industry, law, medicine or other upper middle-class professions, favored constitutional monarchy. Karl Marx (1818–1883), perhaps not appreciated in his political-social originality, the first German sociological thinker (and arguably the second actual Western sociologist after August Comte) wrote from a Social Democratic perspective on Western historical development.

Ranke exhibits the four variables of how inquiry and explanation was pursued in any fourth phase in the arts and the sciences. In an 1833 series of lectures on the "Great Powers" of Europe, he shows (1) a determined set of practice through geo-political elements and tradition; (2) there is a collective view of each nation, where many events generate a general perspective that is more significant than the one event; (3) from that collective view, one achieves a sense of an enduring reality created by the sum of successive events; (4) there is a focus upon material realities rather than a concept-based argument; the material realities can be conceived by the principles by which they affect the individuals collectively. His introduction to the work manifests all four:

> With study and lectures one can comprehend one's perception of history as that of a journey, indeed the successive events of life itself. Even as one event looms large at first, imposing its presence upon us, in time it becomes background, indeed obliterates itself, disappears leaving only an impression that we perceive within a more general view of what has occurred. This general view is all that is accessible to us, increasing over time as part of a sum of

our mental-spiritual possessions. The most prominent moments of comprehensible existence occur in memory and so generate in an enduring manner their living content.

Immediately after a lecture on a significant work, one can see the results in an isolated manner, reviewing the important parts. Yet, it is wise at the same time to see it as part of a sum of more such studies. I am inviting the reader in this present work to learn through some effort the effects of a long historical period, several centuries, and bring them to mind as a contextual whole.

A view of the single moment in history in its truth, in its special development, is of inestimable worth. The particular carries within it a general truth. One can and must raise this single moment into a view of the manifold it represents. The manifold even against our will shows itself as a unity.[1]

Ranke's treatment of France exemplifies this approach:

We start from the assumption that in the sixteenth century freedom in Europe depended upon the opposition and equilibrium between Spain and France. If over powered by one, haven was found in another. As France was weakened and disrupted in this time by internal struggles, a general unhappiness was created among many nations. That is the reason that Henry IV was greeted so warmly as he ended France's anarchy, and recreated the normative European order of things.

Ranke in the above paragraph shows us all four principles of fourth phasal public policy—the customary balance of power that is the gravity that can be disrupted, but finds its equilibrium once more, an imbalance returning to an equilibrium that determines the sequence of events among the nations. The sequence itself is an enduring set of practices, where each must be seen as a segment of a formal causal pattern that endures. Each event is a material reality of persons with distinct policy preferences suitable to the state-of-affairs in pursuing equilibrium with other nations. There are places that must be defended or attacked, and the things of taxes, commerce and the other variables of everyday existence for each particular nation so engaged.

There had been a normative order since the nation-states were formed by 1500, arguably France being the first to have a dynastic family continuity to be fully in control of its business men, the Church, and its own aristocracy. Only the new Protestantism disrupted that between the 1560s and Henry IV. But from 1500 until 1600 we see Spain under the Habsburgs, England under the Tudors, and Russia, under Peter the Great and his Romanov successors finding a balance of power with France. After 1660 France did disrupt that normative balance of power with the strength of its military. The equilibrium of the nations returned with the conclusion of the War of the League of Augsburg from 1689 to 1697, and an even longer-lasting equilibrium at the conclusion of the War of Spanish Succession from 1701 to 1714, ending with the Treaty of Utrecht. Indeed, the peace among European nations established by that treaty lasted until 1740. The

---

1. Leopold von Ranke, *Die grossen Mächte,* http://www.gutenberg.org/files/39669/39669-h/39669-h.htm

atmosphere was one of peace-making and collective thinking. The Abbe St. Pierre spoke of a European Federation.

This equilibrium returned in the decade Ranke was writing, after the French Revolutionary wars, whose disequilibrium causing events were a century after the beginning of the like events between 1689 and 1714. From 1815 through 1854 there were no wars between European countries. Klemens von Metternich had taken leadership at the Congress of Vienna between 1814 and1815 in re-establishing this multi-national equilibrium. It was only the Neo-Absolutism that followed the 1848–1849 revolutions that led to the conflict-filled foreign policies of the now potent manufacturing class.

Ranke wrote the lectures on the "Great Powers of Europe" in 1833, and the perspective came from his own given the politics of nations that surrounded him. He evidently put his trust in monarchical leaders and their prime ministers to main the normative equilibrium that we see as normative only in the fourth phases of a metaparadigm.

Georg Wilhelm Friedrich Hegel (1770–1831)

Hegel wrote in 1815–1816 in his *Proceedings of the Estates Assembly of the Kingdom of Württemberg, 1815–1816* against how the state of Wurttemberg planned a "democratic" constitution, one based solely on age and a minimum of property. His argument then and in later texts was founded upon the idea of some formal competence that made one useful to the state. It was an idea that harked to the rights of ministerials—University graduates who entered the civil service, as well as artisans, and the new industrial leaders whose industries created strength and wealth for the state. Hegel in his arguments will create the idea of certain "estates" of persons within a hierarchy of "estates." This will be a core concept of collective attributes, of the spirit of a fourth phasal understanding. One then became determined by that class membership insofar as the existential possibilities of his or her life. The membership will be enduring, not changing insofar as the organization of the state. Hegel begins this argument in his *Proceedings*:[2]

> Age and property are qualities affecting only the individual himself, not characteristics constituting his worth in the civil order. Such worth he has only on the strength of his office, his position, his skill in craftsmanship which, recognized by his fellow citizens, entitles him accordingly to be described as master of his craft, or has in some other way been accepted into a specific sphere of civil activity. On the other hand, of one who is only twenty-five years old and the owner of real estate that brings him in 200 or more guilders a year, we say "he is nothing". If a constitution nevertheless makes him something, a voter, it grants him a lofty political right without any tie with other civic bodies and introduces in one of the most important matters a situation which has more in common with the democratic, even anarchical, principle of separation than with that of an organic order.

---

2. Hegel, G. (2009).   Review, Proceedings of the Estates Assembly of the Kingdom of Württemberg, 1815–1816 (33 sections). In B. Bowman and A. Speight (Eds.), *Georg Wilhelm Friedrich Hegel: Heidelberg Writings: Journal Publications* (Cambridge Hegel Translations, 32–136). Cambridge: Cambridge University Press. doi:10.1017/CBO9780511596858.004

The great beginnings of internal legal relationships in Germany which presaged the formal construction of the state are to be found in that passage of history where, after the decline of the old royal executive power in the Middle Ages and the dissolution of the whole into atoms, the knights, freemen, monasteries, nobility, merchants, and tradesmen formed themselves into societies and corporations to counteract this state of disorganization. These groups then rubbed against one another for a while until at last they found a tolerable *modus vivendi* neighbours. Since the supreme public authority whose impotence was the direct cause of the need for these corporations was something so loose, these sectional communities forged their bonds of connexion all the more rightly, strictly, even painfully, until they came to constitute a cramping formalism and the spirit of a guild which, because of its aristocratic nature, was a hindrance and a danger to the development of the public authority. After the development of the supreme powers of the state had been completed in recent times, these subordinate communities and guilds were dissolved or at least deprived of their political role and their relation to internal constitutional law. Now, however, it would surely be time, after concentrating hitherto mainly on introducing organization into the circles of higher state authority, to bring lower spheres back again into respect and political significance, and, purged of privileges and wrongs, to incorporate them as an organic structure in the state. A living interrelationship exists only in an articulated whole whose parts themselves form particular subordinate spheres. But, if this is to be achieved, the French abstractions of mere numbers and quanta of property must be finally discarded, or at any rate must no longer be made a dominant qualification or, all over again, the sole condition for exercising one of the most political functions. Atomistic principles of that sort spell, in science as in politics, death to every rational concept, organization, and life. (*Proceedings,* 2009, 262–63)

Hegel several years later takes up the problem of how the lower estates on the hierarchy suffer insofar as the higher estates have more access, given the laws, to more wealth. Here he introduces the concept of "class," based upon wealth. The "concept" is a material reality, not a speculative cause and effect. We will see how Karl Marx takes this element of materially based social fact, and develops it into a theory of history that will be called "historical materialism."

Hegel writes of "class" in his 1820 *Elements of the Philosophy of Right*:

When the activity of civil society is unrestricted, it is occupied internally with *expanding its population and industry*. One the one hand, as association of human beings through their needs is *universalized*, and with it the ways in which means of satisfying these needs are devised and made available, the *accumulation of wealth* increases; for the greatest profit is derived from this twofold universality. But on the other hand, the *specialization* and *limitation* of particular work also increase, as do likewise the *dependence* and *want* of the class [Klasse] which is tied to such work. This in turn leads to an inability to feel and enjoy the wider freedoms, and particularly the spiritual advantages, of civil society.[3]

---

3. G. W. F. Hegel, *Elements of the Philosophy of Right,* ed. Allen W. Wood, trans. H.B. Nisbet (Cambridge: Cambridge University Press, 1991), 266 [Par. 243].

When a large mass of people sinks below the level of a certain standard of living, which auto-matically regulates itself at the level necessary for a member of the society in question, that feeling of right, integrity [*Rechtlichkeit*] and honour which comes from supporting oneself by one's own activity and work is lost. This leads to the creation of a *rabble*, which in turn makes it much easier for disproportionate wealth to be concentrated in a few hands. (Ibid., 1991, 266 [Par. 244])

This vision of "class" in its determinative effects is the collective idea of the fourth phasal address of inquiry and explanation. Hegel does not consider "class" revolution by that "rabble" which Marx will call the "proletariat." Rather he ponders the diffi-culty of solving this problem in differing countries (ibid., 1991, 266–67 note, 267–68 [Pars. 244–246]).

Karl Marx (1818–1883)

Marx begins his thoughts on history and current realities with these observations by Hegel. But, by 1843–1844, he attacks Hegel's limited understanding of the course of history, while seeing the correctness of Hegel's vision of changing historical periods as they have been characterized by certain societal structures of groups of people who have certain degrees of authority. Marx will create the conceptual platform for democ-racy in Europe, seeing how the mercantilistic protections of the lower class, the prole-tariat, with the actions of the monarch and his ministers, addressed poverty and the needs associated with poverty. He will blend these historical economic helpmeets with a leveling of authority learned by the capitalist challenges to mercantilism into a new political-economic set of programs called "socialism." This is the core of his concept, taken from Hegel, of a dialectical address of existent realities. By formulating problems and problem solutions through his taking up new angles and evidence that the challenge and response of the dialectic promises, Social Democracy becomes in this fourth phase a political activism.

What must be borne in mind is that Social Democracy began before Marx insofar as the conceptual ideas of equality and a cooperative society based on an appreciation of how every vocation was equal and necessary given the needs of any group of people who sought mutual subsistence. This idea was born in the thought of Jean-Jacques Rousseau in the first phase of this metaparadigm with his *Social Contract* (1762) and his *Emile* (1762), furthered by Rousseau in the second phase with his *Confessions* (1770), and probed more deeply in concept and action in the third phase by Robespierre, Hebert and Babeuf. The idea of a wholly equal community in its recognition of interdependence was formulated by St. Simon in 1820 and Robert Owen in the same period, followed by Charles Fourier in the 1830s.

Marx knew that solving the problem of whole societal groups could not be accom-plished in small farm communities as experimented with by Robert Owen, who took up St. Simon's ideas in establishing the farm community of New Harmony in the state of Indiana in the United States. New Harmony failed by the late 1820s because of the nor-mative understandings of even some of its members that private property was superior to mutual ownership, especially in the young United States where so much was available. The determinative concept of "communism" came from the strict hierarchies of estates in mercantilism, where by Roman Law, these Estates were under the management of the monarch, and facilitated by a cooperative aristocracy and a civil ministry appointed by the monarch. Communism would be an imposed leveling of all persons whose property was to be in the short term taken by the executive authority, the "dictatorship of the proletariat" that is, an authority that were the people themselves as represented by elected spokespersons. Marx's plan with its several periods—beginning with communism, and progressing to an interdependent socialism, was formulated by his writings, the *Economic and Philosophical Manuscripts of 1844*.

Communism was a phase meant to teach through practice that no one was more important to the civil society than the next person, as all were interdependent—the idea of St. Simon. Once this became a normative state of everyday knowing within the whole society, socialism could then be introduced. Socialism allowed for some individual differences, even as they would be monitored by the "general will" of the social whole. In the socialist phase God will not be outlawed as a perspective among the individuals of a society, but it will be in the prior communist phase. That is because Marx shares the Modernist vision of a final cause wholly in the hands of the human being. Marx writes of this phasal changes in his *Economic and Philosophical Manuscripts of 1844*:

You have been begotten by your father and your mother: therefore in you the mating of two human beings—a species act of human beings—has produced the human being. You see, therefore, that even physically man owes his existence to man. Therefore you must keep sight of the *one* aspect—the *infinite* progression which leads you further to inquire: Who begot my father? Who his grandfather? Etc.

... But since for the socialist man the *entire so-called history of the world* is nothing but the creation of man through human labour, nothing but the emergence of nature for man, so he has the visible, irrefutable proof of his *birth* through his *genesis*. Since the *real existence* of man and nature has become evident in practice, through sense experience, because man has thus become evident for man as the being of nature, and nature for man as the being of man, the question about an *alien* being, about a being above nature and man—a question which implies the admission of the unreality of nature and man—has become impossible in practice. *Atheism* as the denial of this unreality, has no longer any meaning, for atheism is a *negation of God*, and postulates the *existence of man* through this negation; but socialism as socialism no longer stands in need of such a mediation. It proceeds from the *theoretically and practically sensuous consciousness* of man and of nature as the *essence*. Socialism is man's *positive self-consciousness*, no longer mediated through the abolition of religion, just as *real life* is man's positive reality no longer mediated through the abolition of private property, through *communism*. Communism is the position as the negation of the negation, and is hence the *actual* phase necessary for the next stage of historical development in the process of human emancipation and "becoming oneself again" [Wiedergewinnung][4] *Communism* is the necessary form and the dynamic principle

---

4. The translator of the German verbal noun "Wiedergewinnung" mistakenly calls it "rehabilitation," which is from an outside agency. Marx stresses here human agency by the person themselves, which is of this Modernist metaparadigm.

of the immediate future, but communism as such is not the goal of human development, the form of human society.[5]

The historical materialism Marx develops in this 1844 manuscript is more grounded within the material facts of everyday life than Hegel evidences in his writing. Marx will attack Hegel as being "too conceptual" in the manuscript (ibid., 1976, 338–45), Marx centering his own view in how one becomes alienated from oneself through political-social structures by observing the behavior of workers:

> When communist artisans associate with one another, theory, propaganda, etc., is their first end. But at the same time, as a result of this association, they acquire a new need—the need for society (with one another)[6]—and what appears as a means becomes an end. In this practical process the most splendid results are to be observed whenever French socialist workers are seen together. Such things as smoking, drinking, eating, etc., are no longer means of contact or means that bring them together. Company, association, and conversation, which again has society (with one another) as its end, are enough for them; the brotherhood of man is no mere phrase with them, but a fact of life, and the nobility of man shines upon us from their work-hardened bodies. (Ibid., 1976, 313)

I call this insight by Marx "the foundation" of historical materialism because the forms of association between individuals in a society are normative for the tenets of how the individual must relate within that society given how the individuality of self and other is conceived. The ego-based forms of association in a bourgeois society differ from the cooperative stress of mutual understanding in a socialist society. Marx makes this distinction:

---

5. Karl Marx, *Economic and Philosophic Manuscripts of 1844* in Karl Marx and Friedrich Engels, *Collected Works,* Volume 3 (New York: International Publishers, 1976), 305–6. The German edition with the term "Wiedergewinung" is Karl Marx, *Ökonomisch-philosophische Manuskripte* (Frankfurt am Main: Suhrkamp Verlag, 2009), 129.
6. Marx writes "Gesellschaft," which actually is a verbal noun in this passage's usage. Here is the passage in German: "Wenn die kommunistischen *Handwerker* sich vereinen, so gilt ihnen zunächst die Lehre, Propaganda etc. als Zweck. Aber zugleich eignen sie sich dadurch ein neues Bedürfnis, das Bedürfnis der Gesellschaft an, und was als Mittel erscheint, ist zum Zweck geworden. Diese praktische Bewegung kann man in ihren glänzendsten Resultaten anschauen, wenn man sozialistische französische ouvriers vereinigt sieht. Rauchen, Trinken, Essen etc. sind nicht mehr da als Mittel der Verbindung oder als verbindende Mittel. Die Gesellschaft, der Verein, die Unterhaltung, die wieder die Gesellschaft zum Zweck hat, reicht ihnen hin, die Brüderlichkeit der Menschen ist keine Phrase, sondern Wahrheit bei ihnen, und der Adel der Menschheit leuchtet uns aus den von der Arbeit verhärteten Gestalten entgegen." Karl Marx, *Öarl Marx, nen, so gilt ihManuskripte* (Frankfurt am Main: Suhrkamp, 2009), 146. See below for German usage of the equivalent term "vergesellschaften," which Marx used later to express this thought. Marx is here making the argument that the association is at first a principled coming together to politically cooperate as an end, but that intention makes the association itself an end, that is "the need for society with one another."

Yet, since what is needed to address this reality is an understanding of the mutuality between other egoistic individuals, and as the means towards this understanding does not exist yet, every individual must create this mutual context, insofar as they become the joiners of the needs beyond oneself and the objects of these needs. Natural necessity of mutual interests hold the members of bourgeois society together, as estranged as they may seem, so that the bourgeois, not the political life, is the real bond. It is not the state that holds together the atoms of bourgeois society, rather that the atoms are only a conception, the heavenly illusion—in reality a differentiated entity, namely not that of godly egoists, rather egoistic persons.[7]

As we now turn to literature in this fourth phase, from 1820 through the early 1860s, we will see in the short novels of Honoré Balzac how bourgeois society through its egoistic norms which through political-social-economic motives foster that egoism and discourage interdependence, even as the bourgeois norms are reinforced day by day in the conventional manners of daily existence. Marx borrowed the term "bourgeois" from Balzac as Balzac labeled that class of aspiring capitalists with exactly the description that Marx wished to make.[8]

B. Literature

Honoré Balzac (1799–1850)

Balzac devoted 90 novels and novellas between 1829 and 1847 to what he called as a whole "The Human Comedy" "La Comédie Humaine." In order to show the manner in which ordinary conversation in day-to-day meetings affect individuals, Balzac has multiple characters enter several of the novels and novellas, 26 of them, that appear sometimes as protagonists, sometimes as minor figures.[9] Each has his or her own non-learning or learning curves that move idealism to ordinary bourgeois practice. Each can be followed from plot to plot, and one sees how each one affects the others with whom they have social involvements. They serve as synecdoches of the normative social practices of Balzac's France, and even more as he understood them, synecdoches of human nature in its possibilities. Balzac is deeply indebted to the evolutionary research of his time and earlier. His writing was in his mind an analog to what zoologists had discovered and were discovering. What was the nature of the range of thought and behavior in the human species? While writing these novels and novellas, it wasn't until 1842 that he coined the term "La Comédie Humaine" to characterize what he had written. The term was apropos of the comedy of the human species as it sought to survive. He published in that year a collection of the novellas with an Introduction that explained the purpose

7. *Aus dem literarischen Nachlass von Karl Marx, Friedrich Engels und Ferdinand Lassalle,* hrsg. Franz Mehring. Vol. II, *Gesammeltee Schriften von Karl Marx und Friedrich Engels von Juli 1844 bis November 1847* (Stuttgart: J.H.W. Dietz Nachfolger, 1902), 226–27.

8. Sandy Petrey, "The Reality of Representation: Between Marx and Balzac," *Critical Inquiry* 14, no. 3, *The Sociology of Literature* (Spring, 1988), 448–68. Published by: The University of Chicago Press Stable URL: https://www.jstor.org/stable/1343698

9. See https://en.wikipedia.org/wiki/La_Com%C3%A9die_humaine#Recurring_characters.

of his broad title "La Comédie Humaine," and his indebtedness to zoologists for that understanding.

The idea originated in a comparison between Humanity and Animality. He took up the ideas of the Zoologist Geoffroi Saint-Hilaire, using his novels as what he called analogs, to the vision of how Animality evolved according to Saint-Hilaire. Saint-Hilaire (1772–1844), and his mentor Lamarck (1744–1829), saw how the choices of individuals influenced the genetic generation of descendants. Lamarck, writing in the second and third phases of the metaparadigm focused upon individual agency and acquired characteristics, whereas Saint-Hilaire, thriving in the fourth phase of the metaparadigm, stressed how the environ itself shaped the person, and passed on these traits of thought and behavior. Saint-Hilaire turned the taught concepts of Lamarck into his own vector of determinism and durational existence over generations. For Saint-Hilaire, environment and how animals coped, and how humans coped by creating patterns of societal structures, the way commerce and daily needs were housed and distributed, all this was a determinative set of conditions that shaped peoples thought and behaviors.[10] Balzac writes in this 1842 Introduction to Volume One of *La Comédie Humaine*:

> For does not society modify Man, according to the conditions in which he lives and acts, into men as manifold as the species in Zoology? The differences between a soldier, an artisan, a man of business, a lawyer, an idler, a student, a statesman, a merchant, a sailor, a poet, a beggar, a priest, are as great, though not so easy to define as those between the wolf, the lion, the ass, the crow, the shark, the seal, the sheep, etc. Thus social species have always existed, and will always exist, just as there are zoological species. If Buffon could produce a magnificent work by attempting to represent in a book the whole realm of zoology, was there not room for a work of the same kind on society? But the limits set by nature to variations of animals have no existence in society. When Buffon describes the lion, he dismissed the lioness with a few phrases; but in society a wife is not always the female of the male. There may be two perfectly dissimilar beings in one household. The wife of shopkeeper is sometimes worthy of a prince, and the wife of a prince is often worthless compared with the wife of an artisan … Though Leuwenhoek, Swammerdam, Spallanzani, Réamur, Charles Bonnet, Müller, Haller and other patient investigators have shown us how interesting are the habits of animals, those of each kind, are, at least to our eyes, always and in every age alike; whereas the dress, the manners, the speech, the dwelling of a prince, a banker, an artist, a citizen, a priest, and a pauper are absolutely unlike, and change with every phase of civilization …
>
> Some persons, seeing me collect such a mass of facts and paint them as they are, with passion for their motive power, have supposed, but wrongly, that I must belong to the school of Sensualism and Materialism—two aspects of the same thing—Pantheism. But their misapprehension was perhaps justified—or inevitable. I do not share the belief in indefinite progress for society as a whole; I believe in man's improvement in himself. Those who insist on reading in me the intention to consider man as a finished creation are strangely mistaken. *Séraphîta*, the doctrine in action of the Christian Buddha, seems to me an ample answer to this rather heedless accusation.

---

10. https://en.wikipedia.org/wiki/%C3%89tienne_Geoffroy_Saint-Hilaire#Geoffroy's_theory

Hence the work to be written needed a threefold form—men, women, and things; that is to say, persons and the material expression of their minds; man, in short, and life. As we read the dry and discouraging list of events called History, who can have failed to note that the writers of all periods, in Egypt, Persia, Greece, and Rome have forgotten to give us a history of manners? ...

It is no small task to depict the two or three thousand conspicuous type of a period; for this is, in fact, the number presented to us by each generation, and which the *Human Comedy* will require. This crowd of actors, of characters, this multitude of lives, needed a setting—if I may be pardoned the expression, a gallery. Hence the very natural division already known, into the *Scenes of Private Life*, of *Provincial Life*, of *Parisian, Political, Military*, and *Country Life*. Under these six heads are classified all the studies of manners which form the history of society at large, of all its *faits et gestes*, as our ancestors would have said.[11]

The question that seems to be raised by this Introduction is to what degree is the individual shaped by his or her environ "For does not society modify Man, according to the conditions in which he lives and acts." The key word for seeing Balzac's motivations and intentions in this work in the above sentence is "modify."

Balzac believed in the monad of Leibniz, everyone has his or own "monad" or "self" from birth (ibid., 2008, xii). Balzac's own deeper self may be found in his novel *Séraphîta*. The androgyny was discovered by him and owned. This is like Hume's discovery of self, not the invention of self as in Locke. Hume's fourth phasal discernment of the self which is exposed over years of probing memories is what Balzac recognized as "self-improvement" Balzac came to know the presence of the androgyny of his "monad." While it was "modified" by the manners of the society in which he lived, his discernment of its shifting roles in his self-awareness and behavior toward others became a model through the novel *Séraphîta*, and his women throughout his opus, and the men as they related to the women. Balzac generated for his milieu and later an "awareness of how he was aware." He brought Western self-knowledge to a higher level that he hoped his readers might emulate.

All novellas and novels of Balzac begin with a direct description of a house of some kind, and then the character of the people who inhabit it. Balzac shows us the environmental setting, and how social structures attract and condition its recipients, the other services and employments available congruent to the level of society that inhabits such structures. For example, we see an extended description of a shop front with inhabited rooms on the floors above built in the sixteenth century, called the "At the Sign of the Cat and the Racket," as the novel whose plot takes place in the nineteenth century, begins. Each floor is described, evidently structured in different centuries, and symbolizing the character to a degree, of those times:

On rainy morning in the month of March, a young man, carefully wrapped in his cloak, stood under the awning of the shop opposite this old house, which he was studying with the

---

11. Honoré De Balzac, *The Human Comedy, Volume One, At the Sign of the Cat and Racket and Other Works*, trans. Clara Bell and R.S. Scott (Baltimore: Noumena Press, 2008), xii–xiv, xxi and xxiii.

enthusiasm of an antiquary. In point of fact, this relic of the civic life of the sixteenth century offered more than one problem to the consideration of the observer. Each story presented some singularity; on the first floor four tall, narrow windows, close together, were filled as to the lower panes with boards so as to produce the doubtful light by which a clever salesman can ascribe to his goods the color his customers inquire for. The young man seemed very scornful of this part of the house; his eyes had not yet rested on it. The windows of the second floor, where the Venetian blinds were drawn up, revealing little dingy muslin curtains behind the large Bohemian glass panes, did not interest him either. His attention was attracted to the third floor, to the modest sash-frames of wood, so clumsily wrought that they may have found a place in the Museum of Arts and Crafts to illustrate the earlier efforts of French carpentry … . (Ibid., 2008, 3–4)

The text continues with these descriptions of the house's stories for another page. The character of the persons in their limitations now in the early nineteenth century is made through the implicitly through the character of the domiciles in that ancient shop and residences. The same kind of beginning that indicates the sociological constraints of housing, and thereby the general character of its inhabitants, one reads as the onset in his earlier novel *Le Père Goriot* (1835). The novel is dedicated to Saint-Hilaire:[12]

Madame Vauquez, whose maiden name was De Conflans, is an elderly woman who for forty years has kept, in Paris, a family boarding-house situated in the Rue Neuve-Sainte-Genevíéve, between the Latin Quarter and the Faubourg Saint-Marcel. This boarding-house, known as the *Vauquer House* is open to men and women, and to young and old; yet so great its respectability that it has never been assailed by slander. Still, for the last thirty years, no young woman has been seen there, and if a young man should board with Madame Vauquer, it is certain that he receives but a meager allowance from his family. (Ibid., 1960, 1)

Balzac shows in all his novellas and novels how the characters rise above, fall below or sustain the character that is congruent with such housing. Some "modify" through self-discovery, other fail in self-discovery.

C. Biology

Charles Darwin (1809–1882)

It is a fatal fault to reason whilst observing, though so necessary beforehand and so useful afterwards.[13]

Charles Darwin

---

12. Honoré De Balzac, *Père Goriot*, trans. Wallace Fowlie (New York: Holt-Rinehart-Winston, 1960), viii.
13. Charles Darwin, *The Autobiography of Charles Darwin, 1809–1882* (New York: W.W. Norton, 1958), 159.

Herbert Spencer's deductive manner of treating of every subject is wholly opposed to my frame of mind ... over and over again have I said to myself after reading one of his discussions—"Here would be a fine subject for half a dozen years' work".

Charles Darwin (ibid., 1958, 162)

Darwin sought to integrate what had been written about evolution by intense observation over the whole of the fourth phase of this second metaparadigm of Modernism. He knew of his grandfather's work, and lauded his labor of observation (a third phase rigor), but thought him as one who had "an overpowering tendency to theorize and generalize" (ibid., 1958, 153). The work of a third phase inquiry is, indeed, that of amplifying the systematic, theoretic structure of the second phase—where Erasmus Darwin developed his view of botanical and zoological development. Yet, the fourth phase does draw from the work of the earlier theories. As the above first epigraph states, theoretical structure is "so necessary beforehand and so useful afterwards." However, the second above epigraph, criticizing the post-1859 writing of Herbert Spencer—a writing we will turn to in the next metaparadigm's first phase—is suspicious of any conception, generalization or theory that is not carefully founded in material research.

Inquiry into the material facts of existence, being careful not to use the study merely to define a concept, or build an architectonic for future research, attempts to show how existence is determined in an enduring manner. The fourth phases of metaparadigms share this point-of-view. Darwin evidently learned from Montesquieu about how environment conditions life. And, as we have seen, Montesquieu, in the years between 1730 and 1748 did thorough studies of the tongues of sheep to prove in theory of environmental causation.

The long duration of environmental causation for Darwin is strikingly exemplified in his criticism of how in the late-eighteenth century, persisting into his own time, individuals sought to improve the breed of pigeons by selectively mating those who were quicker or had a better homing instinct. Yet, it is the rock pigeon from which these sub-breeds have come, and from which fertility promises their continuance, while those bred domestically, far from the wild of the forest, are not always fertile. Only the "mongrel" pigeon who is bred directly from the rock-pigeon is always fertile. For Darwin this finding supports the significance of what he calls elsewhere the "average" of the species persisting, while the domestically bred sub-species cannot guarantee "the survival of the fittest." Darwin writes:

Great as are the differences between the breeds of the pigeon, I am fully convinced that the common opinion of naturalists is correct, namely, that all are descended from the rock-pigeon (Columba livia), including under this term several geographic races or sub-species, which differ from each other in the most trifling respects.

From these several reasons, namely,—the improbability of man having formerly made seven or eight supposed species of pigeons to breed freely under domestication;—these supposed species being quite unknown in a wild state, and their not having become anywhere feral;—these species presenting certain very abnormal characters, as compared with all other Columbidae, though so like the rock-pigeon in most respects;—the occasional re-appearance of the blue colour and various black marks in all the breeds, both when kept pure and when cross;—and, lastly the mongrel offspring being perfectly fertile, from these several reasons taken together, we may safely conclude that all our domestic breeds are descended from the rock-pigeon or Columba livia with its geographical sub-species.[14]

This theoretical judgment based on extensive study of pigeons, which Darwin sees corroborated by the work of other scientists, will make his judgment markedly different from Herbert Spencer less than a decade later as he champions the most exceptional example of a species in terms of some special performance, rather than a durational endurance of centuries or millennia. When Charles Darwin speaks of new forms of species that arise through the conflict with others of their species and other species, i.e., "the survival of the fittest," he writes:

> Natural selection acts only by the preservation and accumulation of small inherited modifications, each profitable to the preserved being ... A large number of individuals, by giving a better chance within any given period for the appearance of profitable variations, will compensate for a lesser amount of variability in each in in each individual, and is, I believe, a highly important element of success. Though nature grants long periods of time for the work of natural selection, she does not grant an indefinite period; for as all organic beings are striving to seize on each place in the economy of nature, if any one species does not become modified and improved in a corresponding degree with its competitors, it will be exterminated. Unless favorable variations be inherited by some at least of the offspring, nothing can be effected by natural selection ...

> In the case of methodical selection, a breeder selects for some definite object, and if the individuals be allowed freely to intercross, his work will completely fail. But when men, without intending to alter the breed, have a nearly common standard of perfection, and all try to procure and breed from the best animals, improvement slowly but surely follows from this unconscious process of selection. (Ibid., 1979, 132–33)

Darwin clearly defends "nature" versus "nurture." And, as we have seen and will see in our contrast of the phases of the metaparadigm, the first three phases stress the personal agency of nurture, and the fourth phase the determinative effects of nature—be this in politics, literature or biology, as well as in the other arts and sciences.

---

14. Charles Darwin, *The Origin of the Species*, in *Darwin, Texts, Backgrounds, Contemporary Opinion, Critical Essays*, ed. Philip Appleman (New York: Norton Critical Edition, 1979), 109.

Part III

The Third Modern Metaparadigm
c.1860–c.1960

The analytic revelation is a revolutionary force. With it a blithe skepticism has come into the world, a mistrust that unmasks all the schemes and subterfuges of our own souls. Once roused and on the alert, it cannot be put to sleep again. It infiltrates life, undermines its raw naiveté, takes from it the strain of its own ignorance, de-emotionalizes it, as it were, inculcates the taste for understatement, as the English call it—for the deflated rather than the inflated words, for the cult which exerts its influence by moderation, by modesty. Modesty—what a beautiful word! In the German (Bescheidenheit) it originally had to do with knowing and only later got its present meaning; while the Latin word from which the English comes means a way of doing—in short, both together give us almost the sense of the French *savoir faire*—to know how to do. May we hope that this may be the fundamental temper of that more blithely objective and peaceful world which the science of the unconscious may be called to usher in?

Its mingling of the pioneer with the physicianly spirit justifies such a hope. Freud once called his theory of dream "a bit of scientific new-found land won from superstition and mysticism." The word "won" expresses the colonizing spirit and significance of his work. "Where id was, shall be ego," he epigrammatically says. And he calls analysis a cultural labor comparable to the draining of the Zuider Zee. Almost in the end of the traits of venerable man merge into the lineaments of the grey-haired Faust whose spirit urges him:

To shut the imperious sea from the shore away, Set narrower bounds to the broad water's waste.

Then open I to many millions space

Where they may live, no safe-secure, but free

And active. And such a busy swarming I would see

Standing amid free folk on a free soil.

The free folk are the people of a future freed from fear and hate, and ripe for peace.[1]

Thomas Mann, *Freud and the Future* (1936)

---

1. Thomas Mann, "Freud and the Future," in *Essays by Thomas Mann*, trans. H. T. Lowe-Porter (New York: Vintage Books, 1957), 324.

Sigmund Freud is not the only inquirer who set a new level of "being aware of how one is aware" in this third Modern metaparadigm that spanned the years from the 1860s into 1960. However, as Mann indicates in his address to Freud and others on the occasion of Freud's 80th birthday, most call him "the" most paradigmatic mind of his time. Others who share this distinction wholly or to some degree will be either mentioned by me below or the full focus of my treatments of historical thought, literary thought and scientific thought, in the four phases of this metaparadigm.

What was new in this metaparadigm of Modernism insofar as "being aware of how one is aware" is the non-reflective level of immediate consciousness that had been inaccessible in critical self-awareness. This new awareness was the hitherto rarely indicated access and analysis of the limitations in consciousness of how one thought, even with careful reflectiveness upon one's own perspective. This level would require either a stylistic analysis of ordinary discourse in oneself, or, another person or persons who could provide the individual with evidence of aspects of communication that he or she did not see in themselves, or, could not with good faith address. The foundation for this work began with Immanuel Kant in the second Modern metaparadigm with his observations on how a thinker is often unaware of what another can see in his or her statements (*Critique of Pure Reason*, 1965, 310 [A 314. B 370]. And, as well, Kant's challenge to future researchers to comprehend why certain idea formulations reoccur over centuries as a basis for an individual's arguments, without them necessarily seeing this perspective, reflectively, in how they shape evidence (*Critique of Pure Reason*, 1965, 668–69 [A 856, B 884). Important in this pre-third Modern metaparadigm greater awareness of "how one uses language to be aware as one is aware" was Kant's follower, Friedrich Schleiermacher with his work on the use of grammar, which came to be called "stylistics."[2] This taking up of Schleiermacher's insights in the third Modern metaparadigm, whom we will treat, are Freud's teacher, Franz Brentano,[3] Wilhelm Dilthey[4] and Edmund Husserl.[5]

We will speak of Jean-Paul Sartre's idea of "bad faith"[6] in the fourth phase of this metaparadigm. Sartre built upon what Max Scheler had earlier termed this blinding of

---

2. Friedrich Schleiermacher, *Hermeneutics and Criticism and Other Writings*, trans. Andrew W. Bowie (Cambridge: Cambridge University Press, 1998), especially 225–80. See also, Friedrich Schleiermacher, *Hermeneutics, The Handwritten Manuscripts*, trans. James Duke and Jack Forstman (Atlanta, Georgia: Scholars Press, 1977).

3. See Franz Brentano, *Psychology from an Empirical Standpoint*, trans. Antos C. Rancurello, D.B. Terrell and Linda L. McAlister (New York: Humanities Press, 1973), especially 101–76. See also, *The True and the Evident*, trans. Roderick Chisholm, Ilse Politzer and Kurt R. Fischer (London: Routledge and Kegan Paul, 1971), especially 71–73.

4. Wilhelm Dilthey, *Poetry and Experience*, ed. Rudolf A. Makkreel and Frithjof Rodi (Princeton: Princeton University Press, 1985), 229.

5. Edmund Husserl, *Logical Investigations*, trans, J.N. Findlay, Vol. I (London: Routledge and Kegan Paul, 1970), 257–58.

6. Jean-Paul Sartre, *Being and Nothingness*, trans. Hazel E. Barnes (New York: Washington Square Press, 1966), 112–16.

oneself to oneself "ressentiment."[7] Both of these self-denials were other than the insight of Freud, his own teachers, and others, of the even deeper level of the self-construction of non-reflective awareness. The sociological level of self-blindness grew gradually in the second Modern metaparadigm with Rousseau, St. Simon and continued into Karl Marx.

Perhaps the most important foundation for what occurred in historical thought in the initial three phases of the third Modern metaparadigm, from the 1860s through 1920, was the advent of democratic government in all Western nations between 1862 and 1875. The circumstances for that expansion of suffrage to men varied with the countries involved, but there seemed to be a "spirit of the times" that spread throughout the West. Individuality in its new-found complexity became accessible in ideation, but even the less self-aware now knew equality politically. That heightened the meaning of personal individuality. Women responded by beginning their own demand for suffrage in a heightened manner after these male democratic gains.

7. Max Scheler, *Formalism in Ethics and Non-Formal Ethics of Values,* trans. Manfred S. Frings and Roger L. Funk (Evanston, Ill.: Northwestern University Press, 1973), 37, 119, 228–32, 252, 306, 311 and 348.

Chapter Nine

THE FIRST PHASE: SEMINAL IDEATION, C.1860–1870: THE FOCUS UPON DEFINITION AND HYPOTHESIS

A. Philosophy of History

Karl Marx (1818–1883)

In 1857, Marx, always a forerunner of sociological perspectives, addressed his theory of historical materialism from a highly individual perspective. He writes:

> Individuals producing in society—therefore socially determined production by individuals—is naturally he starting point. The single, isolated hunter and fisherman, with whom Smith and Ricardo begin, belongs to the imaginative fancies of eighteenth century Robinsonades [Utopias on the lines of Defoe's *Robinson Crusoë*], which certainly do not, as cultural historians believe, express simply a reaction to over-refinement and a return to a misconceived natural life. As little as Rousseau's *Contrat social*, whereby naturally independent subjects are brought into association and relationship by contract, is (it) based upon such naturalism. This is (not) illusion, the purely aesthetic illusion of small and great Robinsonades. It is, on the contrary, the anticipation of "bourgeois society," which, since the sixteenth century, has been preparing itself for, and, in the eighteenth has made giant strides towards, maturity. In this freely competitive society the individual appears as released from the natural ties, etc. which, in earlier epochs of history made him an appendage of a distinct, limited human conglomerate. For the prophets of the eighteenth century, on whose shoulders Smith and Ricardo still firmly stand, this eighteenth-century individual—the product, on the one hand, of the breaking of feudal social patterns and, on the other, of the new productive powers developed since the sixteenth century—hovers as an ideal of past existence. Not as a historical result but as the starting point of history. Because as Natural Man, conformable to their idea of human nature, not as arising historically but as determined by nature. This illusion has occurred in each successive epoch up till now.

> The further back in history we go, the more does the individual, and thus also the productive individual, appear as dependent, as part of a greater whole: first still quite naturally in the family, and. Expanding thence, in the tribe; later in the various forms of community resulting from opposition between and amalgamation of tribes. Not until the eighteenth century, in "bourgeois society," do various forms of association in society appear to the individual simply as a means to private ends, as external necessity. But the epoch producing this position, that of the individualized individual, is precisely the epoch of the hitherto most highly developed

social (from this point of view, general) conditions. Man, in the most literal sense, is a *zoon politikon,* a political animal, not just a social animal but an animal which can achieve individuation only in society.[1]

Marx in his identification of every individual as needing a community to simply "individuate," that is develop his or her inborn potentialities, is the Leibnizian and Kantian insights. Leibniz with his monad that contains the human community, and Kant, especially in his "Idea for a Universal History from a Cosmopolitan Point-of-View" (1784), capture what Marx will articulate further in this 1857 writing—that is, the "vergesellschaftung," the "sociation" of each person. That is the need for the other as an object of one's own articulation of thought, behavior and the associated skill development.

Marx's 1857 thinking was continued with his 1859 *A Contribution to the Critique of a Political Economy.* There he addressed with ground-breaking, insofar of not only his theory, but the political-social-economic context that would have to be addressed by all future political economists in some fashion. His approach, like all first phasal thinkers, was definitional. One can see the spirit of Hobbes and Rousseau in this work. He defined commodities, the means of exchange, money, and, again with a definitional cogency, the cycle of production, distribution, exchange and consumption. Marx had begun what will be a second phase work, the systematic set of three volumes, *Das Kapital,* which will take up the efforts of the rest of his life, the last volume still being worked on when he died in 1883. This architectonic provided the third and fourth phasal depth of thought that was the height of its effectiveness throughout the Western world.

In an 1867 essay, Marx will write on "value, price, and profit." Marx will focus here upon definitional overviews of how the work associated with each person's articulation of personal skills constrains, perverts and robs that person of the self-value that is potential, and even expressed to the degree allowed by "bourgeois society." Marx understands this expropriation of an individual's own skills as inevitable in the capitalist understanding of individuality, where only those who rise to the level of employing others for personal gain may be free from this fate, condemning others to it because of his or her own needs in that capitalist system. Marx will take the Utilitarian concept of "surplus value," which is the "bourgeois" concept that justifies the profits of the employer of those who do not share the "surplus value," i.e., the profit over the expenditures of the machinery, and most of all, the labor of others. The Utilitarians of the late eighteenth and early nineteenth centuries, Bentham, Malthus, Ricardo, Mill and others, took Adam Smith's idea of the value of labor, and removed it from he or she who labored. Marx cleaved to the implications of Smith's concept of labor value, arguing for each individual to be compensated for their work, but even more, to be able to work in a manner that more fully articulated and enabled them to develop more fully their skills.

---

1. Karl Marx, "From *Grundrisse der Kritik der Politischen Ökonomie (Ground Plan of the Critique of the Political Economy)*", in *The Portable Marx,* ed. Eugene Kamenka (Middlesex, England: Penguin Books, 1983), 375–76.

Marx speaks of the actual degradation of the worker in the current system that makes his own self-value insofar as individuation impossible as he works the machines of the bourgeois owners:

> By selling his laboring power, and he must do so under the present system, the working man makes over to the capitalist the consumption of that power, but within certain rational limits. He sells his laboring power in order to maintain it, apart from its natural wear and tear, but not to destroy it. In selling his laboring power as its daily or weekly value, it is understood that in one day or one week that laboring power shall not be submitted to two days' or two weeks' waste or wear and tear. Take a machine worth £1000. If it is used up in ten years it will add to the value of the commodities in whose production it assists £100 yearly. If it be used up in five years it would add £200 yearly, or the value of its annual wear and tear is in inverse ratio to the time in which it is consumed. But this distinguishes the working man from the machine. Machinery does not wear out exactly in the same ratio in which it is used. Man, on the contrary, decays in a greater ratio than would be visible from the mere numerical addition of work.
>
> In their attempts at reducing the working day to its former rational dimensions, or, where they cannot enforce a legal fixation of a normal working day, at checking overwork by a rise of wages, a rise not only in proportion to its surplus time exacted, but in a greater proportion, working men fulfill only a duty to themselves and their race. They only set limits to the tyrannical usurpation of capital. Time is the room of human development. A man who has no free time to dispose of, whose whole lifetime, apart from the mere physical interruptions by sleep, meals, and so forth, is absorbed by his labour for the capitalist, is less than a beast of burden. He is a mere machine for producing Foreign Wealth, broken in body and brutalized in mind. (Ibid., 1983, pp. 422–23)

The sociological context of the new emphasis upon an individual who comes to realize the aspects of what his understanding of himself and his life are not fully known by him, fits this new effort, here by Marx, to make available to the individual a political-social context, and its debilitating implications for his or her own life. Marx sees what the ordinary individual cannot know in himself or herself because of the normative system of work. Work is that activity always essential to the survival of homo sapiens. Marx sets the possibility of new problem formulations for becoming an individuated individual in his historical problematics of individual self-knowledge as one who necessarily works for his own and his community's survival.

Marx can be seen as a transition mind, writing the 1857 statement on individuality, and the 1867 consideration of the effects of capitalist "surplus value" on the individual. In that transitional sense into the first and then second phase, he keeps focus upon not only the individual, but the role of his or her community context. His later effort to create a systematic architectonic will not take up the individual to this extent, rather explore more fully the economic context of the challenge to a liberated individuality.

The individual we will discuss under the literature of the first phase of this third Modern metaparadigm will similarly be such a transition mind to individuality, speaking at once of both the individual's responsibility to self and to the community of which he is part.

B. Literature/Drama

Henrik Ibsen (1828–1906)

Ibsen wrote two plays in the late 1860s that focused upon the complex situation of a radical individualism within a community. *Brand* was written in 1865, and *Peer Gynt* was written in 1867. Ibsen shows how Brand can take individual leadership in the broader problems of community. This message by Ibsen will be repeated in Nietzsche's 1872 study *Birth of Tragedy* of how Thespis changed Greek drama in 534 BC, becoming the speaker of new truths, ushering in the age of Heraclitus, Parmenides, Aeschylus, Sophocles, Socrates, Plato, Euripides, Aristotle and others. Peer Gynt, on the other hand is a trickster, an Aristophanes at best, who seems valueless to his milieu, which does strive in the spirit of the late nineteenth century to improve. Each protagonist is outside the norms—Brand superior, Peer Gynt inferior. For Ibsen both as individuals have a leadership contribution in their own way.

Ibsen is a transition thinker, for community may be lead, but it is the in-common whole that must be considered insofar as the common good that must be realized. The meaning and outcomes of the dramas insist on that value. Nonetheless, the individual "leadership" is the new metaparadigm's initial phase. Ideas, briefly, albeit well-defined, that will be amplified in the next phases.

Brand is more accessible to a reader of Ibsen's time. He generates a spur of spirit in all who hear him speak. The significance for this first phasal individualism is that focus upon how something is "erlebt" "immediately experienced,", as opposed to a more distanced appreciation of an epic experiential tale that is somewhat removed in the telling. And, as with ancient Greek drama, especially from Aeschylus and Sophocles, that is the audience experience as they probe themselves in the spirit of the dramas, the dramas engaging them wholly as a new way forward in personal interaction and personal choice. In the first act, one who hears Brand speak so characterizes his effect:

Agnes: (Gazes absently before her, and says)
Yes, yes … But tell me did you see …
Einar: What?
Agnes: (Without looking at him, and in a hushed voice, as if in church)
How he grew while he was speaking!
(She goes down along the path. Einar follows her.
A path along the mountain cliff, with a wild precipice out to the right.
Farther off, above the mountain, glimpses of greater heights, with peaks and snow.

What Agnes had listened to was Brand's self-doctrine, for himself, yet for others:

Brand: Nay, 'tis nothing new I aim:
'tis rights eternal I proclaim.
Dogmas and churches come and go;

There is another cause than mine is;
Both saw their first day once, and so
May see their last, for all I know:
Each thing created finds its *finis*,
Gets tainted by the moth and worm,
And must. By changeless nature's norm,
Give place to some yet embryo form,
But there is something that still yet stays,—
The spirit uncreate, once lost,
Then saved, at the Redemption's cost,
In the beginning of the days:
That threw a bridge when faith was fresh,
From Soul, the aspiring Soul, to Flesh.
Now it is hawked about, retail
Since these new views of God prevail.
But from these fragments of the soul,
These spirit-torsos that remain,
These heads and hands, shall spring a Whole,
That God shall know His own again,—
His Man, His Masterpiece, His heir,
His Adam, young, and strong, and fair!

There is still a responsibility to tradition, and that is God, but it is a God who renews, as Nature renews. And, in the spirit of the Modern metaparadigms, it is man who must be that agent, through self-understanding and choice. To bear this message in one's choices, one must be balanced, and not inflated with a self-destructive egoism. Freud and others will offer that warning, but Ibsen knows it, and shows Brand going too far, jeopardizing his own family, because of his lack of penetration to how other isolate leaders can fail to see who they are responsible for, who they are helping, or hurting. At the end of the play, Brand has brought his family into jeopardy upon the snowy mountain to which he has retreated from his community's lack of appreciation for him:

> *In the midst of the great mountain wastes. A storm is gathering, driving the clouds heavily over the snow-fields. Black peaks and mountain-tops stand out now and then, and are veiled again by the mist. Brand is pursuing his way over the fells, bruised and bleeding.*

Brand (stops and gazes back.)
Climbed a thousand from the hollow;
Up the height not one will follow.
Every soul was fain to climb
To a loftier, greater time,—
Spoke some voice in every heart;
Up, and play the warrior's part!
'Tis the sacrifice that chills,

That makes craven of their wills;
Since for all one blood was spilt,
Cowardice no more is guilt!
(He sits down on a stone and looks shrinkingly around him). (Ibid., 1908, 208)

Brand is plunged into self-doubt, and struggles with his own mind and conscience to affirm himself and his choices. He realizes before the end of this act, and the play, that he is not a "Christ" as his one follower, Gerd, would have it. Yet, he also wonders in his last words if his life path was meaningful.

Ibsen was a prophet, much like Rousseau, and did struggle, as Brand, to affirm his self-understanding, which sought an advance in how individuals should affirm their own equality in society, and their own minds. Brand's last words are a question to what he felt as an infinite demand of which he could never be sure:

Brand (crouching before the descending avalanche, cries up):

Answer me, O God above!
In death's jaws: can human will,
Summed, avail no fraction still
Of salvation?—
(Ibid., 223)

Ibsen, through the mind of Brand, does approach modesty—the temper that Thomas Mann felt Freud taught as the new behavior consequent to the new method of "knowing how one was aware." And knowing that, altering the egoism with a critical self-knowledge. Brand still cognizes God, even as he feels Christianity is limited as a belief system that should be surpassed. The surpassing would offer a final cause and formal cause of only self-willed choices, based upon one's inner feeling of rightness in one's intentions. It is this proto-creational secularity that has slowly developed from the 1650s onward. One sees this in Hobbes, in Hume, in Rousseau, in Kant and in Marx, among others. Some see the Divine in a non-proximate manner; others increasingly leave the Divine out of their intention. Ibsen still believes, but takes a position we have heard in Gotthold Ephraim Lessing when in his *The Education of Humankind* he felt a third age where the human shared God's powers was at-hand. Ibsen will in this way further the view and inspire Nietzsche, and later Max Scheler in this presumption.

Ibsen will find a greater modesty in his life and vision. Ibsen's play *Peer Gynt* two years later has a protagonist who does not presume to lead, but only be himself—truly a more modest aim, even as Peer Gynt will have behaviors that must be self-known and ideally altered.

Both plays end with the theme of love. Love is the emotion that levels us as followers or leaders to the ground and understanding of who others are. Love takes time to care, and as Diotima put it in Plato's *Symposium*, it begins with *eros* but rises to *agape*—the love and outreach to others in the non-carnal sense. After these last lines by Brand, a voice from above gives this answer to his questioning plea:

A voice (crying through the thunder roar)

God is love! (Ibid., 223)

The last lines of *Peer Gynt* stress that love is all, the core emotion of human meaning, fulfillment and the gateway to the development which we have called "human problem-solving." Yes, problems are best formulated and addressed through rational knowing, but as they are for the care and progress of humankind in the ways of peace—as even Freud will warrant, even as he qualifies it as a sublimated emotion stemming from egoistic power. Love is the medium in which every context thrives best. We will come to that as we take up group dynamics and encounter groups in the fourth metaparadigm's initial phases.

The last lines of *Peer Gynt* are given to Solveig, a strong, yet caring woman who against all seeming grounds, supports Peer Gynt with love. She sees him as a needy soul who cannot help who he is and what he does. Peer Gynt was rejected by the Devil, as his selfish and foolhardy behaviors were not seen a sinful. He turns to Solveig for succor, for help, as the play ends, we not knowing the final outcome of Peer Gynt, but the end is promising given Solveig's care. She speaks out with Peer Gynt clinging to her, his head in her lap:

Solveig (sings quietly):

Sleep, my love, my own sweet child,
I shall rock thee free from guilt.

The child's safe in its mother's lap.
The lifelong day they play and sleep.

The child's at rest; his mother's breast
Protects him; and in God they're blest.

The boy-child lay, close to my heart,
the live-long day. Now he is tired.

Sleep, my love, my own sweet child,
I have rocked thee free from guilt ...

I have borne thee freed from guilt,
Sleep my love, my own dear child.[2]

The literary critic Harold Bloom of New York University in his book *The Western Canon* has challenged the conventional reading of *Peer Gynt*, stating:

---

2. Henrik Ibsen, *Peer Gynt*, in *Peer Gynt and Brand*, in versions by Geoffrey Hill (England: Penguin Books, 2016), 340–41.

Far more than Goethe's *Faust*, Peer is the one nineteenth-century literary character who has the largeness of the grandest characters of Renaissance imaginings. Dickens, Tolstoy, Stendhal, Hugo, even Balzac have no single figure quite so exuberant, outrageous, vitalistic as Peer Gynt. He merely *seems* initially to be an unlikely candidate for such eminence: *What is he*, we say, *except a kind of Norwegian roaring boy?*—marvelously attractive to women, a kind of bogus poet, a narcissist, absurd self-idolator, a liar, seducer, bombastic self-deceiver. But this is paltry moralizing—all too much like the scholarly chorus that rants against Falstaff. True, Peer, unlike Falstaff, is not a great wit. But in the Yahwistic, biblical sense, Peer the scamp bears the blessing: More life.[3]

I agree with Bloom, to the extent of his self-willing of life, in the Germanic sense of "erleben" "of engaging the immediacy of others sensuously and fully in his interactions. But, this requires, in the mind of Ibsen, a moderation of ego, and a reaching out for the care to be found in love.

C. Biology

Herbert Spencer (1820–1903)

Spencer's major work in the first phase of the third metaparadigm, which would profoundly influence as we will discuss, the idea of one objective truth in the arts and the sciences, was *First Principles of a New System of Philosophy* (1862). He incorporated an earlier publication of 1857 called *Progress: Its Law and Cause* into the foundational structure of his 1862 publication. *First Principles* was much as Thomas Hobbes's *Leviathan* or the initial work in the same phase of the first Modern metaparadigm by Leibniz on logical calculus, definitional, yet an attempt to cover a broad scope of concepts deemed necessary. For these individuals, as well as Spencer, the next phase was a more ordered architectonic of systematic cause and effect of the concepts indicated in the first phase. Spencer's first Preface in 1862 offered the concepts he would define in their significance. The definitions varied in length—as one would find in the earlier treatment of evolution in the first phase of the second metaparadigm of modernism with Caspar Friedrich Wolff.

As I indicated in the epigraph to Charles Darwin's work, he viewed Spencer's thought too concise, being a promise of more research, rather than the measure of Darwin's life work that met the exhaustive standards of the fourth phasal period. He had written in his 1876 *Autobiography*:

> Herbert Spencer's deductive manner of treating of every subject is wholly opposed to my frame of mind ... over and over again have I said to myself after reading one of his discussions—"Here would be a fine subject for half a dozen years' work." (Ibid., 1958, 162)

Spencer himself recognized the terse cogency of definition in the Preface to his 1862 work:

---

3. Harold Bloom, *The Western Canon* (New York: Harcourt, Brace, and Company, 1994), 357.

In anticipation of the obvious criticism that the scheme here sketched out is too exten-sive, it may be remarked that an exhaustive treatment of each topic is not intended; but simply the establishment of *principles*, with such illustrations as are needed to make their bearings fully understood. It may also be pointed out that, besides minor fragments, one large division (*The Principles of Psychology*) is already, in great part, executed. And a further reply is, that impossible though it may prove to execute the whole, yet nothing can be said against an attempt to set forth the First Principles and to carry their applications as far as circumstances permit.[4]

Below is an outline he provided for this work. His treatment of each concept was more ample in the initial volumes. He continued to write into the 1880s on the remaining volumes described.

A SYSTEM OF PHILOSOPHY.

MR. HERBERT SPENCER proposes to issue in periodical parts a connected series of works which he has for several years been preparing. Some conception of the general aim and scope of this series may be gathered from the following Programme.

FIRST PRINCIPLES.

PART I. THE UNKNOWABLE.—Carrying a step further the doctrine put into shape by Hamilton and Mansel; pointing out the various directions in which Science leads to the same conclusions; and showing that in this united belief in an Absolute that transcends not only human knowledge but human conception, lies the only possible reconciliation of Science and Religion.

PART II. LAWS OF THE KNOWABLE.—A statement of the ultimate principles discernible throughout all manifestations of the Absolute—those highest generalizations now being disclosed by Science which are severally true not of one class of phenomena but of *all* classes of phenomena; and which are thus the keys to all classes of phenomena.*

[x]

[*In logical order should here come the application of these First Principles to Inorganic Nature. But this great division it is proposed to pass over: partly because, even without it, the scheme is too extensive: and partly because the interpretation of Organic Nature after the proposed method, is of more immediate importance. The second work of the series will therefore be—*]

THE PRINCIPLES OF BIOLOGY.

Vol. I.

PART I. THE DATA OF BIOLOGY.—Including those general truths of Physics and Chemistry with which rational Biology must set out.

---

4. Herbert Spencer, *First Principles*, Second Edition (London: Williams and Norgate, 1867), xv.

II. The Inductions of Biology.—A statement of the leading generalizations which Naturalists, Physiologists, and Comparative Anatomists, have established.

III. The Evolution of Life.—Concerning the speculation commonly known as "The Development Hypothesis"—its *à priori* and *à posteriori* evidences.

Vol. II.

IV. Morphological Development.—Pointing out the relations that are everywhere traceable between organic forms and the average of the various forces to which they are subject; and seeking in the cumulative effects of such forces a theory of the forms.

V. Physiological Development.—The progressive differentiation of functions similarly traced; and similarly interpreted as consequent upon the exposure of different parts of organisms to different sets of conditions.

VI. The Laws of Multiplication.—Generalizations respecting the rates of reproduction of the various classes of plants and animals; followed by an attempt to show the dependence of these variations upon certain necessary causes.*

[xi]

THE PRINCIPLES OF PSYCHOLOGY.

Vol. I.

Part I. The Data of Psychology.—Treating of the general connexions of Mind and Life and their relations to other modes of the Unknowable.

II. The Inductions of Psychology.—A digest of such generalizations respecting mental phenomena as have already been empirically established.

III. General Synthesis.—A republication, with additional chapters, of the same part in the already-published *Principles of Psychology.*

IV. Special Synthesis.—A republication, with extensive revisions and additions, of the same part, &c. &c.

V. Physical Synthesis.—An attempt to show the manner in which the succession of states of consciousness conforms to a certain fundamental law of nervous action that follows from the First Principles laid down at the outset.

Vol. II.

VI. Special Analysis.—As at present published, but further elaborated by some additional chapters.

VII. General Analysis.—As at present published, with several explanations and additions.

VIII. Corollaries.—Consisting in part of a number of derivative principles which form a necessary introduction to Sociology.*

THE PRINCIPLES OF SOCIOLOGY.

Vol. I.

PART I. THE DATA OF SOCIOLOGY.—A statement of the several sets of factors entering into social phenomena—human ideas and feelings considered in their necessary order of evolution; surrounding natural conditions; and those ever complicating conditions to which Society itself gives origin.

II. THE INDUCTIONS OF SOCIOLOGY.—General facts, structural and functional, as gathered from a survey of Societies and their changes: in [xii]other words, the empirical generalizations that are arrived at by comparing different societies, and successive phases of the same society.

III. POLITICAL ORGANIZATION.—The evolution of governments, general and local, as determined by natural causes; their several types and metamorphoses; their increasing complexity and specialization; and the progressive limitation of their functions.

Vol. II.

IV. ECCLESIASTICAL ORGANIZATION.—Tracing the differentiation of religious government from secular; its successive complications and the multiplication of sects; the growth and continued modification of religious ideas, as caused by advancing knowledge and changing moral character; and the gradual reconciliation of these ideas with the truths of abstract science.

V. CEREMONIAL ORGANIZATION.—The natural history of that third kind of government which, having a common root with the others, and slowly becoming separate from and supplementary to them, serves to regulate the minor actions of life.

VI. INDUSTRIAL ORGANIZATION.—The development of productive and distributive agencies, considered, like the foregoing, in its necessary causes: comprehending not only the progressive division of labour, and the increasing complexity of each industrial agency, but also the successive forms of industrial government as passing through like phases with political government.

Vol. III.

VII. LINGUAL PROGRESS.—The evolution of Languages regarded as a psychological process determined by social conditions.

VIII. INTELLECTUAL PROGRESS.—Treated from the same point of view: including the growth of classifications; the evolution of science out of common knowledge; the advance from qualitive to quantative prevision, from the indefinite to the definite, and from the concrete to the abstract.

IX. ÆSTHETIC PROGRESS.—The Fine Arts similarly dealt with: tracing their gradual differentiation from primitive institutions and from each other; their increasing varieties of development; and their advance in reality of expression and superiority of aim.

X. MORAL PROGRESS.—Exhibiting the genesis of the slow emotional modifications which human nature undergoes in its adaptation to the social state.

[xiii]

XI. The Consensus.—Treating of the necessary interdependence of structures and of functions in each type of society, and in the successive phases of social development.*

THE PRINCIPLES OF MORALITY.

Vol. I.

Part I. The Data of Morality.—Generalizations furnished by Biology, Psychology and Sociology, which underlie a true theory of right living: in other words, the elements of that equilibrium between constitution and conditions of existence, which is at once the moral ideal and the limit towards which we are progressing.

II. The Inductions of Morality.—Those empirically-established rules of human action which are registered as essential laws by all civilized nations: that is to say—the generalizations of expediency.

III. Personal Morals.—The principles of private conduct—physical, intellectual, moral and religious—that follow from the conditions to complete individual life: or, what is the same thing—those modes of private action which must result from the eventual equilibration of internal desires and external needs.

Vol. II.

IV. Justice.—The mutual limitations of men's actions necessitated by their co-existence as units of a society—limitations, the perfect observance of which constitutes that state of equilibrium forming the goal of political progress.

V. Negative Beneficence.—Those secondary limitations, similarly necessitated, which, though less important and not cognizable by law, are yet requisite to prevent mutual destruction of happiness in various indirect ways: in other words—those minor self-restraints dictated by what may be called passive sympathy.

[xiv]

VI. Positive Beneficence.—Comprehending all modes of conduct, dictated by active sympathy, which imply pleasure in giving pleasure—modes of conduct that social adaptation has induced and must render ever more general; and which, in becoming universal, must fill to the full the possible measure of human happiness.*

As one can readily see, Spencer offers a world view that moves through the natural and social sciences, law and morality. What was later referred to by his critics as "Social Darwinism" is an overarching intention throughout this definitional book. "Social Darwinism" will be how the human species orders itself most effectively and efficiently to continually progress in its domination of life on earth. Spencer, in line with the individualistic focus of these initial phases of the metaparadigm, speaks of how among the best competing systems societal reconstruction is effected by the superior members of the society. This social reconstruction which favors the superior performance of members of a nation, in competition with other nations, will benefit the liberation of the best among the nation to contribute to and improve their societal system. The intentional breeding of the best humans so as to realize what he thought was the aim of the existence of the human species was entertained by him by dint of his conversations with the eugenicist, Francis Galton. Spencer picked up in a spiral return the work of Lamarck insofar as acquired characteristics that could be passed on genetically.

An example of his view of the "law of social evolution" in his chapter on the Law of Evolution in Volume I reads:

In the social organism integrative changes are clearly and abundantly exemplified. Uncivilized societies display them when wandering families, such as we see among Bushmen, join into tribes of considerable numbers. A further progress of like nature is everywhere manifested in the subjugation of weaker tribes by stronger ones; and in the sub-ordination of their respective chiefs to the conquering chief. The combinations thus resulting, which, among aboriginal races, are being continually formed and continually broken up, become, among superior races, relatively permanent. If we trace the stages through which our own society, or any adjacent one, has passed, we see this unification from time to time repeated on a larger scale and gaining in stability. The aggregation of juniors and the children of juniors under elders and the children of elders; the consequent establishment of groups of vassals bound to their respective nobles; the subsequent subordination of groups of inferior nobles]to dukes or earls; and the still later growth of the kingly power over dukes and earls; are so many instances of increasing consolidation. This process through which petty tenures are aggregated in feuds, feuds into provinces, provinces into kingdoms, and finally contiguous kingdoms into a single one, slowly completes itself by destroying the original lines of demarcation. And it may be further remarked of the European nations as a whole, that in the tendency to form alliances more or less lasting, in the restraining influences exercised by the several governments over one another, in the system, now becoming customary, of settling international disputes by congresses, as well as in the breaking down of commercial barriers and the increasing facilities of communication, we may trace the beginnings of a European federation—a still larger integration than any now established.

But it is not only in these external unions of groups with groups, and of the compound groups with one another, that the general law is exemplified. It is exemplified also in unions that take place internally, as the groups become more highly organized. There are two orders of these, which may be broadly distinguished as regulative and operative. A civilized society is made unlike a barbarous one by the establishment of regulative classes—governmental, administrative, military, ecclesiastical, legal, &c., which, while they have their several special bonds of union, constituting them sub-classes, are also held together as a general class by a certain community of privileges, of blood, of education, of intercourse. In some societies, fully developed after their particular types, this consolidation into castes, and this union among the upper castes by separation from the lower, eventually grow very decided: to be afterwards rendered less decided, only in cases of social metamorphosis caused by the industrial regime. The integrations that accompany the operative or industrial organization, later in origin, are not merely of this indirect kind, but they are also direct—they show us physical approach. We have integrations consequent on the simple growth of adjacent parts performing like functions; as, for instance, the junction of Manchester with its calico-weaving suburbs. We have other integrations that arise when, out of several places producing a particular commodity, one monopolizing more and more of the business, draws to it masters and workers, and leaves the other places to dwindle; as witness the growth of the Yorkshire cloth-districts at the expense of those in the West of England; or the absorption by Staffordshire of the pottery-manufacture, and the consequent decay of the establishments that once flourished at Derby and elsewhere. We have those more special integrations that arise within the same city; whence result the concentration of publishers in Paternoster Row, of corn-merchants about Mark Lane, of civil engineers in Great George Street, of bankers in the centre of the city. Industrial combinations that consist, not in the approximation or fusion of parts, but in the establishment of common centres of connexion, are exhibited in the Bank clearing-house and the Railway clearing-house. While of yet another species are those unions which bring into relation, the more or less dispersed citizens who are occupied in like ways; as traders are brought by the Exchange, and as are professional men by institutes like those of Civil Engineers, Architects, &c.

At first sight these seem to be the last of our instances. Having followed up the general law to social aggregates, there apparently remain no other aggregates to which it can apply. This however is not true. Among what we have above distinguished as super-organic phenomena, we shall find sundry groups of very remarkable and interesting illustrations. Though evolution of the various products of human activities cannot be said directly to exemplify the integration of matter and dissipation of motion, yet they exemplify it indirectly. For the progress of Language, of Science, and of the Arts, industrial and æsthetic, is an objective register of subjective changes. Alterations of structure in human beings, and concomitant alterations of structure in aggregates of human beings, jointly produce corresponding alterations of structure in all those things which humanity creates. As in the changed impress on the wax, we read a change in the seal; so in the integrations of advancing Language, Science, and Art, we see reflected certain integrations of advancing human structure, individual and social. A section must be devoted to each group. (Ibid., 1867, 316–19 [Par. 111])

One can readily see in the above paragraphs the generality of the concepts without precise examples. Even the example of how Yorkshire in cloth-making has outdone the West of England, how and why are not explained. As Charles Darwin aptly noted, this is an outline for years of further work. Yet, in this first phase of the new metaparadigm, no one ever goes beyond the brilliance of coining key concepts to be later systematized and explored.

Spencer stresses that material reality is always in motion. This vision will undergird indirectly Einstein's theory of relativity, and third phasal discovery of the motion and force of atomic and sub-atomic particles. Other thinkers, such as Ernst Mach in physics, in the initial phases of this metaparadigm will take up this understanding. With Spencer, one can call it a "spiral return" to the hypotheses of Caspar Frederick Wolff, augmenting them with a more micro-focus upon the generation of organic form:

Motion, wherever we can directly trace its genesis, we find to pre-exist as some other mode of force. Our own voluntary acts have always certain sensations of muscular tension as their antecedents. When, as in letting fall a relaxed limb, we are conscious of a bodily movement requiring no effort, the explanation is that the effort was exerted in raising the limb to the position whence it fell. In this case, as in the case of an inanimate body descending to the Earth, the force accumulated by the downward motion is just equal to the force previously expended in the act of elevation. Conversely, Motion that is arrested produces, under different circumstances, heat, electricity, magnetism, light. From the warming of the hands by rubbing them together, up to the ignition of a railway-brake by intense friction—from the lighting of detonating powder by percussion, up to the setting on fire a block of wood by a few blows from a steam-hammer; we have abundant instances in which heat arises as Motion ceases. It is uniformly found, that the heat generated is great in proportion as the Motion lost is great; and that to diminish the arrest of motion, by diminishing the friction, is to diminish the quantity of heat evolved. The production of electricity by Motion is illustrated equally in the boy's experiment with rubbed sealing-wax, in the common electrical machine, and in the apparatus for exciting electricity by the escape of steam. Wherever there is friction between heterogeneous bodies, electrical disturbance is one of the consequences. Magnetism may result from Motion either immediately, as through percussion on iron, or mediately as through electric currents previously generated by Motion. And similarly, Motion may create light; either directly, as in the minute incandescent fragments struck off by violent collisions, or indirectly, as through the electric spark. "Lastly, Motion may be again reproduced by the forces which have emanated from Motion; thus, the divergence of the electrometer, the revolution of the electrical wheel, the deflection of the magnetic needle, are, when resulting from frictional electricity, palpable movements reproduced by the intermediate modes of force, which have themselves been originated by motion." [Ibid., 1867, 197 (Par.66)]

That the forces exhibited in vital actions, vegetal and animal, are similarly derived, is so obvious a deduction from the facts of organic chemistry, that it will meet with ready acceptance from readers acquainted with these facts. Let us note first the physiological generalizations; and then the generalizations which they necessitate.

Plant-life is all directly or indirectly dependant on the heat and light of the sun—directly dependant in the immense majority of plants, and indirectly dependant in plants which, as the fungi, flourish in the dark: since these, growing as they do at the expense of decaying organic matter, mediately draw their forces from the same original source. Each plant owes the carbon and hydrogen of which it mainly consists, to the carbonic acid and water contained in the surrounding air and earth. The carbonic acid and water must, however, be decomposed before their carbon and hydrogen can be assimilated. To overcome the powerful affinities which hold their elements together, requires the expenditure of force; and this force is supplied by the Sun. In what manner the decomposition is effected we do not know. But we know that when, under fit conditions, plants are exposed to the Sun's rays, they give off oxygen and accumulate carbon and hydrogen. In darkness this process ceases. It ceases too when the quantities of light and heat received are greatly reduced, as in winter. Conversely, it is active when the light and heat are great, as in summer. And the like relation is seen in the fact that while plant-life is luxuriant in the tropics, it diminishes in temperate regions, and disappears as we approach the poles. Thus the irresistible inference is, that the forces by which plants abstract the materials of their tissues from surrounding inorganic compounds—the forces by which they grow and carry on their functions, are forces that previously existed as solar radiations.

That animal life is immediately or mediately dependant on vegetal life is a familiar truth; and that, in the main, the processes of animal life are opposite to those of vegetal life is a truth long current among men of science. Chemically considered, vegetal life is chiefly a process of de-oxidation, and animal life chiefly a process of oxidation: chiefly, we must say, because in so far as plants are expenders of force for the purposes of organization, they are oxidizers (as is shown by the exhalation of carbonic acid during the night); and animals, in some of their minor processes, are probably de-oxidizers. But with this qualification, the general truth is that while the plant, decomposing carbonic acid and water and liberating oxygen, builds up the detained carbon and hydrogen (along with a little nitrogen and small quantities of other elements elsewhere obtained) into branches, leaves, and seeds; the animal, consuming these branches, leaves, and seeds, and absorbing oxygen, recomposes carbonic acid and water, together with certain nitrogenous compounds in minor amounts. And while the decomposition effected by the plant, is at the expense of certain forces emanating from the sun, which are employed in overcoming the affinities of carbon and hydrogen for the oxygen united with them; the re-composition effected by the animal, is at the profit of these forces, which are liberated during the combination of such elements. Thus the movements, internal and external, of the animal, are re-appearances in new forms of a power absorbed by the plant under the shape of light and heat. Just as, in the manner above explained, the solar forces expended in raising vapour from the sea's surface, are given out again in the fall of rain and rivers to the same level, and in the accompanying transfer of solid matters; so, the solar forces that in the plant raised certain chemical elements to a condition of unstable equilibrium, are given out again in the actions of the animal during the fall of these elements to a condition of stable equilibrium.

Besides thus tracing a qualitative correlation between these two great orders of organic activity, as well as between both of them and inorganic agencies, we may rudely trace a quantitative correlation. Where vegetal life is abundant, we usually find abundant animal life; and as we advance from torrid to temperate and frigid climates, the two decrease together. Speaking generally, the animals of each class reach a larger size in regions where vegetation is abundant, than in those where it is sparse. And further, there is a tolerably apparent connexion between the quantity of energy which each species of animal expends, and the quantity of force which the nutriment it absorbs gives out during oxidation.

Certain phenomena of development in both plants and animals, illustrate still more directly the ultimate truth enunciated. Pursuing the suggestion made by Mr. Grove, in the first edition of his work on the "Correlation of the Physical Forces," that a connexion probably exists between the forces classed as vital and those classed as physical, Dr. Carpenter has pointed out that such a connexion is clearly exhibited during incubation. The transformation of the unorganized contents of an egg into the organized chick, is altogether a question of heat: withhold heat and the process does not commence; supply heat and it goes on while the temperature is maintained, but ceases when the egg is allowed to cool. The developmental changes can be completed only by keeping the temperature with tolerable constancy at a definite height for a definite time; that is—only by supplying a definite quantity of heat. In the metamorphoses of insects we may discern parallel facts. Experiments show not only that the hatching of their eggs is determined by temperature, but also that the evolution of the pupa into the imago is similarly determined; and may be immensely accelerated or retarded according as heat is artificially supplied or withheld. It will suffice just to add that the germination of plants presents like relations of cause and effect—relations so similar that detail is superfluous.

Thus then the various changes exhibited to us by the organic creation, whether considered as a whole, or in its two great divisions, or in its individual members, conform, so far as we can ascertain, to the general principle. Where, as in the transformation of an egg into a chick, we can investigate the phenomena apart from all complications, we find that the force manifested in the process of organisation, involves expenditure of a pre-existing force. Where it is not, as in the egg or the chrysalis, merely the change of a fixed quantity of matter into a new shape, but where, as in the growing plant or animal, we have an incorporation of matter existing outside, there is still a pre-existing external force at the cost of which this incorporation is effected. And where, as in the higher division of organisms, there remain over and above the forces expended in organization, certain surplus forces expended in movement, these too are indirectly derived from this same pre-existing external force. (Ibid., 1867, 208–10 [Par. 70])

The material cause of evolution is movement, and its effects as force. The idea of constant motion in material reality is understood by philosophers of history in the initial phases of this metaparadigm as the efficient cause of societal structures, commerce, the military and governance. Formal causes for the initial phases of evolution are the processes of movement, force, and the tissues, organ development, and so forth. Phase four material realities, on the other hand, are seen as more enduring, and there is hesitation to change them in any drastic manner. The material causes are external environmental factors, not as much their inner effects within life forms. The efficient causes how the particular species specimens are so affected. The formal causes are how these so determined specimens and species interact with the each other and the conditioning of the environmental factors.

Each understanding has its factual bases, and, each understanding helps us to cope and problem-solve. What each view has in-common is the efforts of the human species to survive and to solve the problems that threaten survival. Knowing of how these phase transpire sequentially can actually help us integrate perspectives of change and duration, individuality and collectivity, to a greater degree. My findings, thus far, suggest that both an individual and an interdependent awareness may be a gain of the Modernism that insists we "become aware of how we are being aware."

Chapter Ten

THE SECOND PHASE: DEVELOPING A SYSTEMATIC STRUCTURE FOR GUIDING NEW INQUIRY AND EXPLANATION C.1870–C.1895

A. Philosophy of History

Friedrich Nietzsche (1844–1900)

Nietzsche's *On the Advantage and Disadvantage of History for Life* (1874) is a guidebook for how an individual should develop his or her own vision of the history in which they stand. Nietzsche's reputation as the calling upon absolute individualism in decision making squares with the understanding of Rousseau, as well as, of course, other thinkers in any initial phases of a metaparadigm. But, the clarity of his formula for constituting a meaningful history for oneself is unparalleled. I will show that formula in Hayden White's spiral return to that theory in our current metaparadigm in his *The Practical Past* (2014), which rather than a menu to follow, carries out this exercise himself, in a third phasal self-inquiry.

There are three steps to Nietzsche's model for a self-formulated history. The first is to select *monumental* events or event-periods that in a chronological line connect you to that ongoing societal history of which you are a part. As a metaphor, one could call this a discernment of a chain of mountains. That is, the second aspect of the model is to go down the slopes of those mountains, discerning a plethora of factual events that explain the reasoning that enables you to understand in a comprehensive fashion why "that mountain" is so vital for the chronological of which you are a part. Nietzsche calls that exercise the *antiquarian,* a practice by some historians that is isolate in itself, simply a pursuit of certain lines of facts without a collocating concept to a larger whole of which it is a member. For example, coin collectors of a period, or, war historians of a period. The third step is coexistent with the initial two, which is to be guided by a *critical* understanding (or theory) in your selection of the mountain chain and the facts that compose its slopes.[1]

---

1. Friedrich Nietzsche, *The Advantage and Disadvantage of History for Life,* trans. Peter Preuss (Indianapolis, Ind.: Hackett Publishing Company, 1980), 14–22.

Armed with this knowledge, one is prepared, in Nietzsche's understanding, to change history as one sees his or her duty insofar as they themselves are "agents" of history. He writes:

> … we require history for life and action, not for the smug avoiding of life and action. Only so far as history serves life will we serve it. (Ibid., 1980, 7)

Nietzsche will pose a certain state of mind that fosters being an agent of history guided by one's own understanding of the correct vector, given the self-construction with the three conceptual categories of *the monumental, the antiquarian and the critical.* As Luther did when called before the Holy Roman Emperor, Charles V, at the Diet of Worms in 1521, he asked the Emperor for a night's rest and a second day so that he could organize his thoughts. Then, on the second day, facing the Emperor, he stated: "Here I stand, I cannot do otherwise. Unless I am convinced by the testimony of the Scriptures and by clear reason (for I do not trust in the pope or councils alone, since it is well known that they have often erred and contradicted themselves), I am bound by the Scriptures I have quoted. My conscience is captive to the Word of God. I cannot and I will not retract anything, since it is neither safe nor right to go against conscience. May God help me. Amen."[2] Luther had clarified his self-understanding of what history it was within which he stood, and was clear about what action to take. That action was not in that moment merely a citation of his historical knowledge, rather an act whose vector was clear. As Nietzsche said of such a moment:

> Whoever cannot settle on the threshold of the moment forgetful of the whole past, whoever is incapable of standing on a point like a goddess of victory without vertigo or fear, will never know what happiness is, and worse yet, will never do anything to make others happy. Take as an extreme example a man who possesses no trace of the power to forget, who is condemned everywhere to see becoming. Such a one no longer believes in his own existence, no longer believes in himself; he sees everything flow apart in mobile points and loses himself in the stream of becoming. He will, like the true pupil of Heraclitus, hardly dare in the end to lift a finger. (Ibid., 1980, 9–10)

Later in the text, he calls this Heraclitus-like man, "superhistorical" in his posture and knowing (ibid., 12). What Nietzsche means is that having reflected, centering his mind upon the justifications of an intended action, in that moment of action, there is an openness to the circumstances, and, an intuitive comprehension of what is to be said and done. This is Luther in the moment of the second day, knowing how to address Charles V, and the others in that room. We will see this process in Martin Heidegger's existentialism in the mid-twentieth century, as well as in his student, Jean-Paul Sartre. Knowing what is right can be complemented by knowing how the other(s) who are addressed "are aware." This added knowledge is especially relevant in fourth phasal knowing and behavior. Luther's genius was in his Early Modern metaparadigm, fourth phasal. Nietzsche, himself, was more of a Rousseau—not adept at moving others personally.

---

2. https://blog.cph.org/read/everyday-faith/lutheranism/did-luther-really-say-here-i-stand

Nietzsche's architectonic as an intention for others is spoken to in a direct manner later in his text, as he emphasizes the needs for a critical set of conceptions to guide one in the three steps of a self-composed history. He writes of the difficulty of any exact history, but after this reflection, speaks of its necessity. He reflects on the Austrian writer Franz Grillparzer:

> Grillparzer goes on to say; "what else is history, after all, than the way in which the spirit of man apprehends what for him are *impenetrable* events; unites elements of which God only knows whether they belong together; replaces the unintelligible with something intelligible; introduces its concepts of externally oriented purpose into a whole which surely admits only purposes with an inner orientation; and again assumes chance where a thousand little causes were at work. Every person at the same time has his individual necessity so that millions run in directions parallel to each other in crooked and straight lines, cross, support and restrict each other, strive forward and backward and in this assume the character of chance for each other, and so, leaving out of account the influences of natural event, make it impossible to demonstrate a penetrating all-inclusive necessity of events." Just such; a necessity, however, is to be brought to light as a result of that "objective" view of things! (Ibid., 1980, 35)

For Nietzsche, this being oriented and then acting was the ethical necessity in this new time of democracy, as it was instituted in all the West between 1861 and 1875. Each of us must act insofar as our own understanding, and from this, in his augmentation to Franz Grillparzer's skepticism, a self-generating order of human society and increasing progress in human understanding would result. The last paragraph of his book tells us that he has created a parable of each of us to follow as we deem fit, even as he has constructed the lineaments of that parable for us:

> This is a parable for each one of us: he must organize the chaos within himself by reflecting on his genuine needs. His honesty, his sound and truthful character must at some time rebel against secondhand thought, secondhand learning and imitation. Then, he will begin to comprehend that culture can be something other still than *decoration of life*, that is, fundamentally always only dissimulation and disguise. For all adornment hides what it adorns. Thus the Greek concept of culture—in contrast to the Romance concept—will be unveiled to him. The concept of a culture as a new and improved nature, without inside and outside, without dissimulation and convention, of culture as the accord of life, thought, appearing, and willing. Thus he will learn from his own experience that it was through higher strength of *ethical* nature that the Greeks achieved a victory over all other cultures and that every increase of truthfulness must also be a preparatory advancement of *true* culture. Even if this truthfulness may on occasion seriously harm the notion of culture which just then enjoys respect, even if it occasions the fall of an entire decorative culture. (Ibid., 1980, 64)

It must be borne in mind that Nietzsche understood the Übermensch as a person in transition.[3] It is not being superior or beyond human, rather a person who develops

---

3. See Karl Jaspers, *Nietzsche, An Introduction to the Understanding of His Philosophical Activity*, trans. Charles K. Wallraff and Frederick J. Schmitz (Baltimore: Johns Hopkins University Press, 1997), 167–68.

self-consciousness, consciousness of the other, and effective behaviors that transcend what has been normative. This understanding can be traced at least back to Diderot in 1753 when he spoke of constant development in species.[4]

B. Literature/Drama

Henrik Ibsen (1828–1906)

Ibsen's *Pillars of Society* (1877) is an architectonic of concepts whose interrelationship enables a more complex exploration of each of them in his later plays. The overarching concept in this play is of a "pillar" of society. One can see a classical reference that almost certainly prompted his work in Proverbs 9, 1, where the seven pillars of wisdom are introduced as a guiding principle: "Wisdom hath builded her house, she hath hewn out her seven pillars."[5] Ibsen is constructing a model for the major pillars of such a house—that will be human societies. What follows in Proverbs 1 through 31 is the seven pillars and their converse. The pillars are the major solutions for the self-blindness in its many forms that work against a sound society. The pillars are essentially wise human agents who are able to address through language and behavior the resistance to a better way among others, who are not wise. In the spirit of all Modern metaparadigms, there is an effort by Ibsen to instruct others into a new level of awareness of how they must be aware now in modern society. His use of the Book of Proverbs in the Bible is a striking discovery by him, for the Book of Proverbs gives evidence of how those who "know how they know and do not know" can become wise. Ibsen's drama is quite sociological in its concern with communal norms that improve the interdependent behavior of its members. But, it stresses individual insight, decision making, and the bold action in the light of what one discerns that one sees in Nietzsche's *The Advantage and Disadvantage of History for Life*.

Each of Ibsen's characters is either a pillar and/or its converse, in several ways. The generic concept of a pillar in Proverbs is "wisdom." All human beings are capable of wisdom, even after a life that shuns it. The main character in this play is a person who becomes "wise" after 15 years of deceit. Karsten Bernick carries much of the meaning of *Pillars of Society*. He can be seen throughout almost the entire four acts of the play as the archetype of what in Proverbs 10, line 9 reads: "He that walketh uprightly walketh surely, but he that perverteth his ways shall be known" (ibid., 1967, 678). He will finally expose his own deceit in the final act, only by listening to several of the wise persons who he had heretofore not allowed to affect him. Then, in the midst of a tragic vortex that is unfolding because of his actions, tragic to his family, but also to the whole community, he learns to "walketh surely." The 15 years of deceit enabled him to represent his entire community as what it meant to be "a pillar of society," but this meaning was what Nietzsche called a "decorative" set of norms "—*decoration of life*, that is, fundamentally always only

---

4. Denis Diderot, *Thoughts on the Interpretation of Nature and Other Philosophical* Works, trans. Lorna Sandler (Manchester: Clinamen Press, 1999), 73, 75–76.
5. Proverbs in *The New Scofield Study Bible, Authorized King James Version* (New York: Oxford University Press, 1967), 678.

dissimulation and disguise. For all adornment hides what it adorns" (Nietzsche 1980, 64). Almost every individual in Karsten Bernick's community contributes to "a decoration of life" whereby each feels better than those of lower economic classes, and do nothing to improve, in the sense of becoming wise as Proverbs would have it, either themselves or the others with whom they interact.

Wisdom sees our ongoing equality with everyone. Wisdom does not elevate oneself, rather it compels us to "love" the other by helping them to see who they can be, who they "naturally" are. Proverbs 10, line 12 "Hatred stirreth up strifes, but love coverth all sins" is applicable to three women who are keystones to the averting of the tragic ending, and whose love finally generates wisdom in Karsten Bernick. One of these women is Lona Hessel, the older half-sister of Bernick's wife, Betty Bernick, neé Tønnesen. Lona defines the false notion of the "pillars of society" represented by the deceitful Karsten Bernick, calling it essentially what Nietzsche later will call it, a *"decoration of life."* The core dialogue over the meaning of this converse of a "pillar of society" occurs in Lona's conversation with Bernick in Act Three:

Bernick: Look into anyone's heart, and you'll find every individual has at least one dark secret to hide.
Lona: And you call yourselves pillars of society!
Bernick: Society provides none better.
Lona: Then why go on supporting a society like this? What does it live by? Lies and pretenses—and nothing else. Here you are, the town's leading citizen, well-off, happy, enjoying power and honor—you, who have branded an innocent man a criminal.
Bernick: Don't you think I'm deeply aware of my injustice to him? And don't you think I am ready to make good on it?
Lona: How? By confessing it?
Bernick: Is *that* what you want!
Lona: What else could right that kind of injustice?
Bernick: I'm rich, Lona. Anything Johann asks for—[6]

Lona demands honesty, forthrightness, and not what Bernick and another central self-deceiver, the schoolmaster, Rørlund, will call "moderate demands" upon one's fellows. Bernick states: "People have to learn to moderate their demands on each other if they're going to make their mark in the community they live in. Betty had gradually grown to realize that, which is why our house has become a model for our fellow citizens" (ibid., 1978, 65). Rørlund had earlier said: "When a man's called forth to be morally, a pillar of the society he lives in, then he can't be careful enough" (ibid., 1978, 30).

One of the "pillar" messages of the play, carried by Lona and three other women, is to act as Nietzsche would later more dramatically state, "standing on a point like a

---

6. Henrik Ibsen, *Pillars of Society* (1877) in *Henrik Ibsen, The Complete Major Prose Plays,* trans. Rolf Fjelde (New York: Plume Books, 1978), 76–77.

goddess of victory without vertigo or fear" (ibid., 1980, 9). Being careful is anathema to a "pillar" response. Dina characterizes how one should always act: "be natural" (ibid., 1978, 53). To say and do what is in one's heart, without hesitation, without fear for oneself or for the other, would they incorporate wise advice, is being beyond the emotion of fear, which holds us back from a "natural" or "true" act. Fear is the hesitation to enter wisdom's structure. Being natural is being a "natural" agent for change in the circumstances in which one finds oneself, and throughout the play, that is in the company of people who need that wisdom.

When Lona first arrives in the community, suffocated by the pseudo "respectability" of the women in her half-sister Betty's house, she is asked by Rørlund, who wants her to leave, "… what can *you* do for *our* society?" Lona answers: "I can air it out—Reverend" (ibid., 1978, 39).

The "respectable ladies" in their shunning of anyone outside their circle, hesitating to leave the protection of the rooms in which they meet, are an evident violation of the "pillar" which might be called "acquaint oneself with all others, as you and they are all equal." Proverbs 9, 13–18 reads:

A foolish woman *is* clamorous; *she is* simple, and knoweth nothing. For she sitteth at the door of her house, on a seat in the high places of the city, To call those who pass by, who go right on their way, Whoso *is* simple, let him turn in hither; and *as for* him that lacketh understanding, she saith to him, Stolen waters are sweet, and bread *eaten* in secret is pleasant. But he knoweth not that the dead *are* there, and *that* her guests are in the depths of sheol (Hades). (Ibid., 1967, 678)

Ibsen seems to have used these several verses to plan the first scene in Act 1. This is the democracy championed by Ibsen, in Brand, and now in *Pillars of Society*. The United States of America is that site where it flourishes for Ibsen throughout this play.

Another core "pillar" is "love," that medium in which Brand concluded his life as the voice from above gave him the answer he sought. Love is central to the conclusion of the play as Bernick finally accepts the wisdom of his family, averting tragedy to its members. Lona surprises him by saying she came back from America, not for vengeance, but for her care concerning him, he who had rejected her for money, and spread rumors concerning her half-brother.

Lona: I didn't come back to expose you. I came back to shake you up so you'd speak out of your own free will. (Ibid., 1978, 104)

Her care for the women of the small "respectable" circle, wanting to "air out" their way of living and thinking, could be called *agape*. Her particular focus on Bernick is somewhere between that and an earlier *eros*, "care" nonetheless. Bernick feels that with her presence, and admits to himself that his leadership of the respectable norms as they are is simply having been "a puppet of society, no more than that" (ibid., 1978, 102). Some critics wonder why Ibsen allows Bernick to have a happy ending going forward in his life. The answer to that is Ibsen's demand of self and others that they love each other in the several ways love has been defined. Ibsen's sense of forms of love is that of Diotima in Plato's *Symposium*. Martha Bernick, Karsten's sister, has loved Johann, Lona's half-brother, for 15 years at a distance, but on his return, and his new-found love for Dina Dorf, she opens her heart to him and understands why she can't be loved by him (ibid., 1978, 100).

How the characters of wisdom speak to each other, and speak to the unwise, is another pillar. The form of converse can be seen in Proverbs, for example, Proverbs 17, 1: "Better *is* a dry morsel, and quietness therewith, than an house full of sacrifices, *with* strife". The "modesty" of the wise women, such as Martha, Dina, who quietly sought to preserve her natural self, never creating a scene with those who tried to control her, and even Lona, who came to shake things up, is evident. Lona in her confrontation with Bernick assures him she will destroy the letters he had sent her 15 years before, even after having married Betty, Lona's half-sister for only her money, where he admitted guilt in allowing Lona's half-brother be scandalized for a reputed embezzlement. This is the modesty of which Thomas Mann spoke in reference to Freud—one of the epigraphs for this metaparadigm. Freud will speak with this voice, and Ibsen, long before Freud, knew only this modesty could enable love to thrive between disputants. To have this modesty is to have not only self-restraint, based upon "knowing how one is aware, and how that awareness can be altered through willed reorientation," but it is a self-willed finding of that reorientation through often extended reflection. Rare individuals have mastered that series of self-change toward others in history, but now, in the third metaparadigm of Modernism, it became a public standard to be achieved. We will see this most particularly in the fourth phase, c.1920 into the 1950s. Yes, a period of the rise of hate-based doctrines, and a world war, but also, a time when leaders all over the world sought to act otherwise, and new models of how to interact became a new micro-sociology.

The pillars I have identified in Ibsen's *Pillars of Society*—acting with the courage that comes from knowing what is natural for you, not fearing any consequences, acting with love, and in that acting modestly—will become the standard for literary protagonists by the end of this this third metaparadigm of modernism. It will be based upon greater self-definition and the seeing of others accurately, but with empathy. In this second phase, these issues and concepts will be merely defined through glimpses in the characters of literature. An architectonic of how to live in this manner toward self and others is created in this play. Later dramas, will amplify this self-struggle among those who interact.

C. Psychology

Franz Brentano (1838–1917)

Brentano writes in his 1874 text *Psychology from an Empirical Standpoint* of the core contribution he will make to the science of psychology in this second phase of the third Modern metaparadigm. In his work, he picks up concepts that enable him to augment, in a spiral fashion, the "being aware of how one is aware" that was initially formulated in the literature of the 1650s by Anton Ulrich, where intention was not always fully understood, initially, and the intentions of self and others were gradually discerned. Somehow, literary inventiveness throughout the metaparadigms were a step ahead of the probing of human consciousness, even as other disciplines contributed in their own manner to the significance of how "being aware of how one was aware, and how others were aware as well." Chladenius, for example, in the third phase of the first Modern metaparadigm had recognized that significant questions when historical interpretation of an event was involved, differed among individuals by dint of their singular modes of thought. Chladenius was spurred to this inquiry by Leibniz before him. Gradually, all disciplines became capable of contributing in an augmenting manner how to more deeply discern the character of a person's viewpoint, and, even, as in this case, why and how the viewpoint may differ within the same person, let alone other persons. Brentano was the teacher of Freud. And, as we will see, Brentano's brilliant discoveries laid the groundwork of the discoveries of his student, Sigmund Freud. Perhaps, most succinctly Brentano writes in his chapter "A Survey of the Attempts to Classify Mental Phenomena" of the history of what he pursues in his inquiries, citing Aristotle:

> Another classification given by Aristotle divides mental phenomena … into *thought* and *desire* … in their broadest sense … In the class of thought, in fact, Aristotle includes not only the highest activities of the intellect, such as abstraction, making universal judgments, and scientific inference, but also sense perception, imagination, memory and expectation based on experience. In the class of desire we find both high aspirations and strivings as well as the lowest drives, and along with them all feelings and affective sates—in short, all mental phenomena which are not included in the first group.

> If we inquire why Aristotle united in this classification those phenomena that he had separated in his first classification, it is easy to see that he was led to do this by a certain similarity between sensual and appearance, and intellectual and conceptual presentation and affirmation. The same similarity is apparent between the lower appetites and higher aspirations. To use an expression we have already borrowed from Scholastics, he discovered the same mode of *intentional inexistence* in both cases.

> [fn. This expression had been misunderstood in that some people thought it had to do with intention and the pursuit of a goal. In view of this, I might have done better to avoid it altogether. Instead of the term "intentional" the Scholastics often used the expression "objective." This has to do with the fact that something is an object for the mentally active subject, and, as such, is present in some manner in his consciousness, whether it is merely thought of or also desired, shunned, etc. I preferred the expression "intentional" because I thought there would be an even greater danger of being misunderstood if I had described the object of thought as "objectively existing," for modern-day thinkers use this expression to refer to what really exists as opposed to "mere subjective appearances."]

> On the basis of this principle, (Aristotle) put activities that had been united by his first classification into different classes. For the reference to the object is different in thought and in desire. And it is precisely this that differentiates the two classes for Aristotle. He did not believe that they are directed toward different objects, but that they are directed toward the same object in different ways. He said clearly, both in his treatise *On the Soul* and in his *Metaphysics* that thought and desire have the same object. It is first present in the faculty of thought and there the desire stirs [cf. *De Anima*, III, 10. *Metaphysics*, XII, 7]. So the first classification was based upon differences in the bearers of mental phenomena and the extent to which they

are distributed over a wider or narrower range of beings endowed with mental faculties; the second is based upon differences in their mode of reference to the immanent object. The order of the succession of the member classes is determined by the relative independence of the phenomena. Presentations belong to the first class, but a presentation is the necessary antecedent condition of any desire.[7]

Brentano will explore these two classes of "intentional" thought which Aristotle discerned, and in doing that formulate what we have come to understand through Freud as the "unconscious" formulation of perspectives. Brentano, however, will consider the unreflective formulation of an object or state-of-affairs in our attention to the immediacy of our experiential moment as of our consciousness, albeit a spontaneous seeing, sensing, and knowing of what we can then reflectively judge. Indeed, the initial judgment is non-reflective and thus what Freud will call "unconscious." That is, how we depict through figural and verbal determination the object or state-of-affairs immediately, before we reflectively interact with it. Freud will combine the two Aristotelian classes of seeing, sensing and judging, and acting toward, which Brentano will also do, but to which Aristotle only referred. That will be the human gain with the thinkers of this third Modern metaparadigm—how we think and judge prior to reflective choice with countless observed examples.

Brentano's concept can be called "non-reflective consciousness" or "non-conscious consciousness," which Freud later simplifies into the concept of that which is "unconscious" in our mental processing. "Intentional inexistence" says the same thing: what one intends is formulating an object or a state-of-affairs solely in consciousness, albeit a consciousness inaccessible initially, and, with Freud even until much later, that is thus "inexistent" insofar it is wholly in one's mind, one's consciousness. Persons, place and things of the objective environ may occasion this immediate consciousness, but the formulation of such presences is wholly in thought. Though it may be in the environment, it is how we know that of the environment. As Kant said in the *Critique of Pure Reason* "Thoughts without content are empty, intuitions without concepts are blind" (1965, A 51, B 76).

Brentano also speaks of the non-conscious consciousness as "inner consciousness." In exploring this in more depth in his subsequent writing of the third phase, i.e., his 1904 "The Equivocal Use of the term 'Existent'," he will explore how we formulate sentences at this level of non-consciousness, each formulation expressing the viewpoint, and in this sense, the emotional intention of the one who judges.[8] For example, in this later work he shows his contact with stylistics, in the deep structural sense of how sentences are formed (ibid., 1971, 655–66). Inner consciousness formulates what becomes spoken or written in two steps: "Every mental act, ... is accompanied by a

---

7. Franz Brentano, *Psychology from an Empirical Standpoint*, ed. Oskar Kraus, trans. Antos C. Rancurello, D.B. Terrell, and Linda L. McAlister (New York: Humanities Press, 1973), 180–81.
8. Franz Brentano, "The Equivocal Use of the Term 'Existent'" in *The True and the Evident*, ed. Oskar Kraus and Roderick M. Chisholm, trans. Roderick Chisholm, Ilse Politzer and Kurt R. Fischer (London: Routledge and Kegan Paul, 1971), 65–73.

twofold inner consciousness, by a presentation which refers to it and a judgment which refers to it, the so-called inner perception, which is an immediate, evident cognition of the act" (*Psychology from an Empirical Standpoint* 1973, 143). He then adds a third kind of inner consciousness: "Experience shows that there exist in us not only a presentation and a judgment, but frequently a third kind of consciousness of the mental act, namely a feeling which refers to this act, pleasure or displeasure which we feel toward this act" (ibid., 1973, 143).

Thus, Brentano opens a wholly new dimension to "be aware of how one is aware," but this dimension can only be accessed by special study into the fact sentences and even figural expression is immediate, wholly formed before our ability to reflect upon them. Training in the recognition and interpretation of these artifacts of mind can enable us to discern more deeply of "how we are aware." This is a step in human problem formulation and problem solving that makes cooperation more possible, even while it makes the human being a more complex species than heretofore understood at this depth. That Aristotle had inklings and insights speaks of the lost classical knowledge, which in itself is a caution in our sense of constant progress.

Chapter Eleven

THE THIRD PHASE: MATERIAL INQUIRY INTO THE VERIFIABILITY OF SPECIFIC CONCEPTS, AND CONFLICT OVER THE IMPLICATIONS OF THE FINDINGS C.1890–C.1920

A. History

Heinrich Friedjung (1851–1920)

Heinrich Friedjung's *The Struggle for Supremacy in Germany 1859–1866* was published in 1897. It is the first history to contain "oral history," published interviews, as Appendixes to the main text, of three key figures within the years covered. In that, Friedjung augments the significance of differing ways of comprehending in-common events articulated as a basis in historical knowledge by Chladenius in the 1740s. Chladenius did not give living examples, but hypothesized certain ways the same event might be seen differently by observers. In his writings, Chladenius also speculated, based upon Leibniz's understanding of each individual having a differing logic of seeing, the idea of each individual having an inherent "Sehepunkt," "point-of view" as they offered their understanding of the course of an event. A century after Chladenius, Karl Marx used the idea of how events were perceived historically from a sociological argument of differing normative socioeconomic views. This was done more thoroughly than Chladenius, but still within the framework of the kinds of generalizations made by Chladenius. Marx's use of empirical arguments was an augmentative step. Indeed, one could call it a "spiral return" to the fourth phase and century before, with Marx's addition of empirical support for the differing ways a capitalist or a mercantilist, or, a socialist as himself, might see something such as the revolutions of 1848.

However, Friedjung's 1897 use of the words of individuals to support the differing views of the struggle for supremacy in Germany and Austrian-Germany from 1859 to 1866 was a greater insight into how an individual thought and acted. In that, Friedjung took the concept of Chladenius and Marx, which was one perceived events from their adherence to a political-social or economic philosophy, and re-applied as an individual understanding. This shift in the concept from the general to the individual corresponds to what Thomas Kuhn saw in the phases of a metaparadigm, where concepts were re-oriented to match the phasal norms of the metaparadigm. Friedjung writing in the third phase had taken up what Chladenius and Marx wrote in the fourth phase. For Chladenius and Marx, one's views were determined by the general, normative views of their political-social understanding. For Friedjung, what was stated by the participants, were wholly an individual point-of-view, articulated in depth by the word for word transcription.

This can be understood as the beginning of oral history. Interestingly, Franz Boas (1858–1942), writing in the same phasal period, created the discipline of cultural anthropology. Boas interviewed various native peoples, attending their oral accounts of their history, and using these oral transcripts to structure an accurate idea of what their culture had been and now was. For example, in 1888, Boas published his research among the Kwakiutl Indians of British Columbia. Using diverse accounts gathered, he identified key words to show what was in-common among the statements of his respondees, and generated not only a history of these people, but their in-common belief system.[1]

Max Weber (1864–1920), another contemporary of Friedjung, was one of the founders of the new discipline of sociology (albeit, the actual beginnings with August Comte and Karl Marx in the fourth phase of the second Modern metaparadigm of the period 1820 to 1860). Weber used oral interviews in his sociological inquiries. Weber was known to have said there is no "average" person, and the abstraction of statistics was of little value. Weber, in his studies of productivity, was famous for creating open-ended questions in factories in order to see the complex differences in individual workers who were affected by their engagement with the tools of manufacture.[2]

Among the biographies of Max Weber we read of his focus upon individual differences, and how these individuals while each differing to a degree, nonetheless can create general change in institutions from their agentive action. A Nietzsche-like position, and certainly Weber read Nietzsche.

> The principle of methodological individualism, which holds that social scientists should seek to understand collectivities (e.g. nations, cultures, governments, churches, corporations, etc.) solely as the result and the context of the actions of individual persons, can be traced to Weber, particularly to the first chapter of Economy and Society, in which he argues that only individuals "can be treated as agents in a course of subjectively understandable action." In other words, Weber contended that social phenomena can be understood scientifically only to the extent that they are captured by models of the behaviour of purposeful individuals—models that Weber called "ideal types"—from which actual historical events necessarily deviate due to accidental and irrational factors. The analytical constructs of an ideal type never exist in reality, but provide objective benchmarks against which real-life constructs can be measured.[3]

Heinrich Friedjung most probably read Nietzsche. Friedjung was a close associate with Viktor Adler, the founder of the Austrian Social Democratic Party in 1889. Adler named

---

1. https://repository.si.edu/bitstream/handle/10088/13090/USNMP-11_709_1889. pdf?sequence=1&isAllowed=y
2. Max Weber. "Zur Psychophysik der industriellen Arbeit," Part II, in Archiv für Sozialwissenschaft und Sozialpolitik, XXVIII Band, 1. Heft (January, 1909), for Weber's recognition that responsible analysis of work include firsthand reports from workers of their state-of-mind and the motives for their actions.
3. https://en.wikipedia.org/wiki/Max_Weber

his son, Friedrich, after Nietzsche, and he himself wore a "Nietzsche moustache," the vogue of the time. Friedjung, besides being what was recognized then as the leading Austrian historian, was also active in political-social issues.[4] The idea of oral interviews for Friedjung seemed a genuine extension of his political activity. He wanted to know why certain individuals made the choices they did in Prussian- German and Austrian-German foreign policy in the years that encompassed his history. Friedjung was an activist, even as he was a careful, evidence-based historian. He believed his historical work would help shape future relations between the new Germany and Austria-Hungary.

In his interview with Otto von Bismarck, Friedjung is given Bismarck's view of a decades-long perspective he had toward a possible war with Austria, which he had written of since the 1850s:

> Prince Bismarck granted the author an interview on June 13, 1890. The conversation turned on the negotiations during the visit of Bismarck and King William to Schönbrunn in the autumn of 1864, and I asked the Prince whether Austria had been at all inclined to cede Schleswig-Holstein to Prussia in return for a guarantee of her Italian possessions. The Prince replied:

> "I cannot remember any such Austrian offer; and I do not think, as far as I can rely only on my memory, that one was made. But according to my intentions then and later, we could easily have agreed to it; for a firm alliance with Austria was always my aim, and my royal master would have gladly given such a guarantee in order to acquire Schleswig-Holstein and remain friendly and at peace with Austria."

> "The Austrian statesmen," I remarked, "seem to have regarded the joint possession of Schleswig-Holstein as more important, for the position of Austria as a great power, than Milan which was already lost and could not be regained."

> "I do not wish," the Prince replied, "to criticize, but merely to give an account of what happened. Rechberg was not opposed to such a solution and I had been on good terms with him ever since our time at Frankfort. I always believed that it would be necessary to reach an understanding, an alliance with Austria, but I only succeeded in this much later, in 1879." …

> Here I interrupted that the Prince's dispatches from Frankfort clearly showed that he already regarded war as the necessary method of solving the German question.

> "In general, certainly," was the answer, "but not all the time, not in the day-to-day incidents of policy. It would be a misinterpretation of the spirit of politics to believe that a statesman can formulate a comprehensive plan and decide what he is going to do in one, two, or three years. Schleswig-Holstein was certainly worth a war; but you cannot pursue a plan blindly. You can only give a general indication of your aim—the statesman is like a man wandering in a forest who knows his general direction, but not the exact point at which he will emerge from the wood. It was difficult to avoid a war with Austria, but he who is responsible for the

---

4. See Heinrich Friedjung, *The Struggle for Supremacy in Germany 1859–1866*, trans. A.J.P. Taylor and W.L. McElwee (London: Macmillan and Co., 1935), xiii–xiv. For Viktor Adler's adherence to Nietzsche, see Julius Braunthal, *Viktor und Friedrich Adler, Zwei Generationen Arbeitrebewegung* (Vienna: Wienervolksbuchhandlung, 1965), 28–29.

lives of millions will shrink from a war until all other means have been exhausted. (Ibid., 1935, 313–14)

Friedjung interviews Count Rechberg, and Rechberg's account of his meeting with Bismarck in 1864, corroborates Bismarck's account. Thus, we see how oral interviews give more depth to evidence, even while situating an in-common event in each person's own understanding:

"In the Schleswig-Holstein question I kept to the firm ground of treaties; even for reasons of our own domestic policy, I could not allow that the principle of nationality alone should decide. I regarded the Prussian alliance as the best for Austria. It is true that we did not make a sufficiently definite agreement as to what was to happen to the Duchies after the war, but things happened more quickly than I wanted. At the end of the war we took the Duchies from Denmark and I wanted Schleswig to go to Prussia and Holstein to Austria, in the hope of getting a guarantee for Venice in exchange for Holstein later on. Biegeleben was of a different opinion and agreed to the surrender of Schleswig-Holstein jointly to Austrian and Prussia, because otherwise, he said, 'we should have no point of conflict with Prussia'."

"My general aim was to maintain friendly relations with Prussia, because I did not believe that Austria was strong enough to wage war. This was what was on my mind during the negotiations with Bismarck at Vienna in the summer of 1864. I said to him one evening that Austria and Prussia ought to remain friendly, for then not a shot could be fired in Europe without their permission. 'I agree,' Bismarck said, 'but our domestic situation is inevitably driving us towards war abroad.' 'Napoleon', I replied, 'is in the same position and as far as I can see he will not be able to keep the opposition quiet for long without a successful war. Then there will be war in Europe and we can beat France.' Bismarck agreed to my suggestion that we should reach an agreement for the event of a French attack. It was midnight; I hurried over to Biegeleben and instructed him to draw up a draft treaty on these lines. Biegeleben, however, refused, so I drew up the treaty myself and it was approved by the two sovereigns he next day." (Ibid., 1935, 318–19)

Reading closely, one can see that there was an understanding between Rechberg and Bismarck in 1864. However, Bismarck was informed by Rechberg of a perceived weakness in Austria insofar as military preparedness. Rechberg, even later, did not see how this enabled Bismarck "to come out of the forest" in 1866. The long range vision of Bismarck is indirectly substantiated by how Rechberg recalls the conversation. Rechberg, on the other hand, never saw the absolute need for conflict. We learn a great deal from oral histories that otherwise we could not see clearly from merely the artifacts of treaties, letters or other indirect means. Friedjung has, indeed, furthered the deepening of "how we are aware of how others are aware."

B. Literature

Thomas Mann (1875–1955)

Thomas Mann's *Death in Venice* will present to us the in-depth story of an individual seeking to change how they view themselves and others. The complexity of the changes toward what Mann will see as a realization of modesty and peace with others, rather than the assertive, conflict-filled perspective of the protagonist's earlier vision of himself and the world, will fulfill what Ibsen strived for in his delineation of what must be addressed in his *Pillars of Society*. Ibsen achieved this complexity in his later plays in such works as *A Doll House* (1879) and *An Enemy of the People* (1882). Living and acting on one's life principle is there in full, but also a way to accommodate the reality of others. Mann is chosen here because his short story, which has been called a novella, the form of short novel developed in the eighteenth century by Goethe and others, is of the new genre of the short story that became so vital in Western culture in the 1890s. That is a story more focused upon one life issue, giving it the treatment of a third phasal explication of one or more concepts outlined as essential to a new view of human nature by the second phase of thought. We will see the use of the "case history," Freud's contribution to literature, as we look at psychology in this third phase. Depth in one issue and its associated complexities generates the short story. The third phase will find many conflicting visions of similar issues as human nature is redefined with both the new and the revisiting of the traditional as modernist progress in comprehending the human species.

Mann's protagonist will be tracked in his shift in personality toward a new way of thought and behavior, without sacrificing his earlier principle of thinking and acting true to principle, as an autonomous change agent. The protagonist of *Death in Venice* is Gustav von Aschenbach, a writer. Before the onset of the story, Aschenbach had written stories of assertive leadership, the life of the Frederick the Great, the individual imposing his will upon a whole people. The key principle and moral of life, quoting Frederick the Great's favorite motto "durchhalten," "hold fast."[5] Throughout the story this motto will never be relinquished by Aschenbach. Yet, Aschenbach's life will be one the models what Mann had written about Freud in the epigraph to this metaparadigm of Modernism, one must find love and modesty through self-transformation. "Aschenbach" can be translated as "ashes in the stream." What one was will be over in the course of this story. If one can address what has been, transforming it to what can be, then is, one has directly or indirectly adhered to the process of this third phasal, in depth, moral of self-examination and change. Life moves on, it changes—a point-of-view of all held in all initial phases of any metaparadigm. The narrator has a friend of Aschenbach state clearly the change in him at the beginning of the story:

> "You see, Aschenbach has always lived like this"—here the speaker closed the fingers of his left hand to a fist—"never like this"—and he let his open hand hang relaxed from the back of his chair. It was apt. (Ibid., 1989, 9)

The last paragraph of the story portrays Aschenbach collapsed in his beach chair, having attended his "love," the Polish youth Tadzio, standing at a distance in front of him, gesturing out to sea (ibid., 1989, 73). His "relaxed" torso had crossed the threshold into a

---

5. Thomas Mann, "Death in Venice," in *Death in Venice and Seven Other Stories*, trans. H.T. Lowe Porter (New York: Vintage Books, 1989), 9.

life he had begun as his infatuation of Tadzio began in Venice. The transformation that began, and ostensibly would continue into another life, is affirmed by Mann's reference to Plato's *Phaedrus,* a critical reference by him in the story, offered by Aschenbach. As in the *Phaedrus*, where Socrates breaks his normal pattern, and travels outside of Athens to a stream in the neighboring countryside, attracted to and attending Phaedrus, Mann's Aschenbach will brave an outbreak of cholera in Venice to remain in sight of his beloved Tadzio on the beach. Aschenbach's lesson to himself and the reader with this quotation from the *Phaedrus* reads:

> "For mark you, Phaedrus, beauty alone is both divine and visible; and so it is the sense way, the artist's way, little Phaedrus, to the spirit ... For you know that we poets cannot walk the way of beauty without Eros as our companion and guide. We may be heroic after our fashion, disciplined warriors of our craft ... (Ibid., 1989, 70–71)

To fully appreciate what it means to follow Eros, and how this leads to not only one's spirit, but a self-transformation of the spirit, Mann certainly expects his reader to go to not only the full story of the *Phaedrus,* but perhaps even more importantly, to Plato's *Symposium,* in which Phaedrus is once more instructed. And, therein Diotima's description of how love moves from Eros to Agape in the process of one's life experience, if one is to know love in its manifold ways. Especially, to know it as something that brings modesty and peace among individuals:

> And so when his prescribed devotion to boyish beauties has carried our candidate so far that the universal beauty dawns upon his inward sight, he is almost within reach of his final revelation. And this is the way, the only way, he must approach, or be led toward, the sanctuary of Love. Starting from individual beauties, the quest for the universal beauty must find him ever mounting the heavenly ladder, stepping from rung to rung—that is, from two to *every* lovely body, from bodily beauty to the beauty of institutions ... [6]

The "ladder of love" is the step-by-step access to the movement from Eros to Agape, as well. It is how we "love" others with a modesty that enables us to learn how to interact within the conventional norms and institutions of the state to change them from within, or without, if needs be (seeing "without" as non-violent resistance, in figures such as Gandhi and Martin Luther King). Mann's protagonist falls in love with Tadzio who is Polish. This is an implicit message of the story that national competitions and hatred are a disease that interferes with the loving of one's fellows. Poland was hated by Germans in their competition over the commerce of Eastern Germany. In 1908, the historian Friedrich Meinecke had written a philosophy of history entitled *Cosmopolitanism and the Nation-State,* cautioning Germans against what he saw as an uncharacteristic chauvinism, given their normative cosmopolitanism of the past, by dint of the Holy Roman Empire.

---

6. See Plato, *Symposium* in *The Collected Dialogues of Plato, Including the Letters,* ed. Edith Hamilton and Huntington Cairns, trans. Michael Joyce, Bollingen Series LXXI (Princeton: Princeton University Press, 1961), 562–63 [211].

The cholera plague can be seen as symbolic of the normative sickness that led to World War I. *Death in Venice* is written in 1911, and all of Europe awaited a coming war with a heightened xenophobic nationalism in each country as normative. Mann will have a background disease of the mind in his treatment of fascism in a 1929 short story called *Mario and the Magician.*

As the story ends, Mann has awakened our attention to the rank of a soul, by encouraging us to read the *Phaedrus.* Plato lists the value of souls in this dialogue, with a lover who seeks wisdom or beauty. The poet is only sixth in the hierarchy of souls, as he is imitative. The king and warrior, as Frederick the Great, who seeks justice among others, is second.[7] Mann will adhere to "holding fast" to his intentional openness and work toward changing his direction of personal agency, as his soul is ready to be reborn as a lover in the best sense. Mann was not an agnostic. However, he was a keen psychological observer of persons and cultures. As an individual, he did change beliefs over the years, shedding his cultural anti-Semitism as he attended Freud's writings, and considered the Jewish background of his wife, and thus, his several children by her. I see him as iconic for the message of the third Modern metaparadigm, which saw the possibility for true equality among individuals and nations based upon the self-transformative learning of our hidden obstacles to openness to others, and in that to self-actualization of all our human capabilities.

C. Psychology

Sigmund Freud (1856–1939)

Freud's case studies were written in the third phase of this third Modern metaparadigm. Each went into the complexities of a particular "unconscious" set of memories that affected normal behavior. While we do not have "self-therapy" in a strict sense, Freud always believed that his patients cured themselves, albeit with his facilitating presence. The process was a rigorous, continual exercise of reflective judgments. Only a strong personal agency in the patient could lead to an in-depth understanding of his own everyday understandings and behavior. The person must not only be reflectively aware of what is present to all in a situation of mutual intercourse, but also cognizant of one's own unseen intentions, that new augmentation human consciousness.[8]

Thus his case studies are of individuals who sought to widen and deepen self-understanding into particular issues in all the complex depth and detail of a third phase

---

7. See Plato, *Phaedrus* in *The Collected Dialogues of Plato, Including the Letters,* ed. Edith Hamilton and Huntington Cairns, trans. R. Hackworth, Bollingen Series LXXI (Princeton: Princeton University Press, 1961), 495 [248 d].

8. Freud writes in one of his last essays that his discovery has enlarged what mentality in humans must now be understood. Everyday judgment is an integration of what has been known as "consciousness," but now must include what is the so-called "unconscious." See Sigmund Freud, "Some Elementary Lessons in Psycho-Analysis (1938)," in *Sigmund Freud, Collected Papers,* Vol. V, ed. James Strachey (New York: Basic Books, 1959), 378–79.

study. Freud's study of "The Rat Man," is that of a study of what he terms "obsessional neurosis." Written in 1909, the study is of a young man who recently graduated the University.

Freud begins the case study with what he says was his only condition for treatment "namely, to say everything that came into his head, even if it was *unpleasant* to him, or seemed *unimportant* or *irrelevant* or *senseless*. I then gave him leave to start his communications with any subject he pleased..... ."[9] Freud emphasizes in his treatment that the patient leads with his or her own selection of topics. Indeed, in the largely monologue that proceeds, the physician must curb his curiosity in order to allow this inner agency of the patient to shape the subject he or she wishes to address:

> The true technique of psychoanalysis requires the physician to suppress his curiosity and leaves the patient complete freedom in choosing the order in which topics shall succeed each other during the treatment. At the fourth sitting, accordingly, I received the patient with the question: "And how do you intend to proceed today?" "I have decided to tell you something which I consider most important and which has tormented me from the very first." He then told me at great length the story of the last illness of his father ... (Ibid., 1963, 33–34)

The patient felt guilt in that he was not present at the death. Here Freud inserts into the conversation a concept of interpretation about "guilt." In the initial phases of a metaparadigm, concepts play a larger part in their instrumental application to an audience, be it a patient or a reader, than the deterministic materialism cited in fourth phases. There is more certainty in inquirers in fourth phases, whereas in the initial phases concepts are often but guides, and often used as organizers for diverse factual discoveries. Freud interjects a thought as the patient speaks of the recurrent emotional torment of his guilt, and he brings up the psychoanalytic concept of mistaken association because of a repression of memory:

> He told me that the only thing that kept him going at that time had been the consolation given him by his friend, who had always brushed his self-reproaches aside on the ground that they were grossly exaggerated. Hearing this, I took the opportunity of giving him a first glance at the underlying principles of psychoanalytic therapy. When there is a *mésalliance*, I began, between an affect and its ideational content (in this instance, between the intensity of the self-reproach and the occasion for it), a layman will say that the affect is too great for the occasion—that it is exaggerated—and consequently the inference following the self-reproach (the inference, that is, that the patient is a criminal) is false. On the contrary, the physician says: "No. The affect is justified. The sense of guilt cannot in itself be further criticized. But it belongs to another content, which is unknown (*unconscious*), and which requires to be looked for. The known ideational content has only got into its actual position owing to a mistaken association." (Ibid., 1963, 35)

---

9. Sigmund Freud, "Notes Upon a Case of Obsessional Neurosis (1909)" in *Three Case Histories*, ed. Philip Rieff (New York: Collier Books, 1963), 20.

Freud had by the time the case studies were written, actually in what for other was already the third phase, by the late 1890s, developed a systematic theory of concepts that became "psycho-analysis." Freud was an exception in moving through the initial two phases as he was born in 1856, and received his training in psychiatry in the 1880s. His case is worth contemplating, as the effect of writing in a certain phasal period is generated by the thought of a time. Freud was pressed to come up with a theory, but it required a decade—thus, his in-depth studies began later in the third phase of his cultural time than more mature minds in other disciplines. In the third phasal reality, selected concepts of a systematic theory are given more careful study. Freud began those studies after 1905, and continued them into the early post-World War I years, when, as I will detail, he turned to collective issues of societal normality and abnormality.

Above, as he explains to his patient the theory of repressed memory and its false associations of an affect, he has begun to probe more deeply what he understood as the "core" concept of psychoanalysis, and that is "repression."[10] This is an interesting concept as it is one of the instances indicated by Thomas Kuhn, where a new inquirer took a concept from the fourth phase of the previous metaparadigm, and changed its meaning. Freud's use of "repression" stressed the role of an unconscious intentional thought by a patient, rather than a physically determined condition of the nervous system, where certain energies are repressed chemically, that the psychiatry that Freud had inherited had employed as. Gustav Fechner's psychophysics promoted that idea in the 1850s. He saw dream imagery as being an indication of the material problems of the person's physical system. Interestingly, Johan Friedrich Herbart (1776–1841), writing in the third phase of the previous metaparadigm, held what became Freud's position, that of force of an unconscious mental process conditioning the physical energy of the individual (ibid., 1979, 66–67).[11] Thus, the "spiral return" of ideas of certain phases, and the transformation of an established concept between the initial three phases and the fourth phase.

10. Frank J. Sulloway, *Freud, Biologist of the Mind, Beyond the Psychoanalytic Legend* (New York: Basic Books, 1979), 123–26.
11. For a thorough discussion of the concept of "repression" and its changing understanding, see Frank J. Sulloway, *Freud, Biologist of the Mind, Beyond the Psychoanalytic Legend* (New York: Basic Books, 1979), 66–67, 123–26, 196–97 and 367–75.

Chapter Twelve

THE FOURTH PHASE: INTEGRATING THE NEW FOUR CAUSAL UNDERSTANDINGS WITH THE TRADITIONAL C.1920–C.1960

Now if we turn to a more customary perspective and view our society through the eyes of Earth's inhabitants, we discover that the functioning or malfunctioning of groups is recognized increasingly as one of society's major problems. In business, government, and the military, there is a great interest in improving the productivity of groups. Many thoughtful people are alarmed by the apparent weakening and disintegration of the family. Educators are coming to believe that that they cannot carry out their responsibilities fully unless they understand better how the classroom functions as a social group. Those concerned with social welfare are diligently seeking ways to reduce intergroup conflicts between labor and management and among religious and ethnic groups. The operation of juvenile gangs is a most troublesome obstacle in attempts to prevent crime. It is becoming clear that much mental illness derives in some way from the individual's relations with groups and that groups may be used effectively in mental therapy …

When in the late 1930s group dynamics began to emerge as an identifiable field the empiricist rebellion was well along in social psychology and sociology, and from the outset group dynamics could employ the research methods characteristic of empirical science. In fact, group dynamics is to be distinguished from its intellectual predecessors primarily by its basic reliance on careful observation, quantification, measurement, and experimentation.

But one should not identify group dynamics too closely with extreme empiricism. Even in its earliest days, work in group dynamics displayed an interest in the construction of theory and derivation of testable hypotheses from theory, and it has come progressively to maintain a close interplay between data collection and the advancement of theory.

Dorwin Cartwright and Alvin Zander (1953)[1]

We have here to deal only with the psychological influence of the environment. This does not mean that the somatic effects of environment, for example, of nutrition or climate, do not have great psychological significance. On the contrary, the somatic as well as the psychological influence of the environment is constantly operating on the entire child.

---

1. Dorwin Cartwright and Alvin Zander, eds., *Group Dynamics, Research and Theory*, 3rd edition (New York: Harper & Row, 1968), 3–5.

It has long been recognized that the psychological influence of the environment on the behavior and development of the child is extremely important. Actually, all aspects of the child's behavior, hence instinctive and voluntary behavior, play, emotion, speech, expression, are codetermined by the existing environment.

Kurt Lewin, *Environmental Forces in Child Behavior and Development* (1935)[2]

The fourth phase of this third Modern metaparadigm is essentially still with us in the fourth Modern metaparadigm in which we now think (having arguably begun c.2020). What we study in this chapter as new breakthroughs will become "tradition" in our present. As I complete this text, we will leave our discussion in the third phase of the fourth Modern metaparadigm (c.1970–c.2060). What will come has already had its inception in thought. This very text can be considered as a threshold of the fourth phase of the fourth Modern metaparadigm, as it studies the configurational criteria in a comparative manner of all inquiry and explanation since the mid-1640s to the present. My thesis includes the assertion that the existing environment of thought conditions how inquiry is conceived, how problems are formulated, how inquiry is explained, as well as how problems then are addressed and answered. My argument is to integrate the value of approaches that stress individualism, autonomy, chance and the insight into what has not yet been perceived, albeit a spiral return to these orientations, with the value of collective, interdependent models of inquiry and explanation, and the determined aspects that brook little free will and autonomy.

Social psychology, group dynamics and self-critical autobiographies began as normative fourth phasal inquiries after World War I. While the environmental conditions themselves fostered this, as the war plunged thought within nations to consider how to interrelate with other nations, for example, the establishment of the League of Nations, historically, this was a period experienced in every previous metaparadigm. This was the phase for integrating the old and the new, the phase for understanding how groups were constructed and interacted, the phase for comprehending how humans were conditioned by both their social and natural environments.

What was new in this fourth phase of the third Modern metaparadigm, beginning in the 1920s, was a closer understanding of how the human mind is conditioned by unconscious thinking, which in itself is conditioned by the interactive experience with others. This led to advances in psychiatry with Freud and other depth psychologists, the creation of a fields of sociology that were considered social psychology, and also, "microsociology," which I will explore as the discipline of group dynamics. In literature, autobiography now linked its self-vision to the environment, as a change agent linked to others of like-mind in altering the conditioning factors of the existing political-social norms. Norman Mailer's *Advertisements for Myself* (1959) being an example, but a host of others, including autobiographies or autobiographically based works coming out of

2. Kurt Lewin, *Environmental Forces in Child Behavior and Development* in *A Dynamic Theory of Personality, Selected Papers,* trans. Donald K. Adams and Karl E. Zener (New York: McGraw Hill Book Company, 1935), 66.

England, France, Germany and other nations. The key element is in formulating one's intentions within what has conditioned them beyond one's immediate knowing.

A. History/Philosophy of History

Sigmund Freud (1856–1939)

Freud's *Civilization and Its Discontents* (1930) generated a new vision of how societal norms functioned. Its conceptual lessons helped explain the growing fascism of his immediate culture. His theory rested on the bases he had developed in his individual psychology in the pre-World War I decades. He took the concept of the individual "super-ego" and argued that a nation's normative thought, and even their institutions could become a pathological superego, much as an individual could generate a pathological superego.

A moral imperative entered his argument. Freud's humanism was a philosophy, not the factual evidence of his work as a psychiatrist. But, every morality is a reflective set of judgments upon life experience, not a set of discerned material facts, rather our choice of understanding. Such an understanding can be rigorous, looking at actions and their consequences. Freud's realization of how "unconsciously" an individual erects pathological responses to everyday events enabled him to see factual evidence in how the laws of a society and their other institutions could compel pathological responses if one was to obey the written and unspoken laws of a society. Or, how a society could constitute the best in the normative reflective behaviors they supported. Why did normative pathologies arise in a culture? Freud, as a Jew, had a lifetime of knowing how they arose.

Freud is not the only investigator of the pathological normative or healthy normative in the third and now the fourth phases of this third Modern metaparadigm. Since democracy, the will to do better than the best was present in all disciplines, especially those of human thought and behavior. Interesting is the parallel to Freud's concept of "repression" in the third phase of this Modern metaparadigm in the discipline of phenomenological philosophy, a field that arose with Freud's own teacher, Franz Brentano, and those influenced also by Brentano, as also Edmund Husserl and Max Scheler. Max Scheler developed a concept called "Ressentiment," which can loosely be translated as our English-speaking concept of "resentment," but with Scheler it was a powerful, often unconscious attitude, often misdirected, to harm another stimulated by the enduring resentment that can become an act of revenge.[3] In 1926 Scheler had fashioned the earlier 1912 discussion of individual "Ressentiment" into that of how a society could have norms that exercised this harm to others, based upon delayed, often repressed impulses. Scheler's 1926 text is considered an extensive argument of the "sociology of knowledge."[4]

---

3. See Max Scheler, *Über Ressentiment und moralische Urteil* (Leipzig: Wilhelm Engelmann, 1912), especially 4–5, where the initial characterization is given.
4. See Max Scheler, *Die Wissensformen und die Gesellschaft* (Leipzig: Der Neue-Geist Verlag, 1926). For a discussion of the text, see Howard Becker and Helmust Otto Dahlke, "Max Scheler's Sociology of Knowledge," in *Philosophy and Phenomenological Research* 2, no. 3 (March 1942): 310–22.

He shows how differing national norms are constructed within the institutions and the laws, and how individuals are educated. A society could be guided by "Ressentiment," in its norms, as cultural norms arise in the interaction with other societies. In this movement from the individual to the societal, there is an indisputable foreseeing of Freud's argument. Did Freud read Scheler? Probably not. Yet Freud is notorious for not giving credit. He never gave written credit to his teacher Franz Brentano for Brentano's revelation of non-conscious consciousness.[5]

Regardless, Freud's concept of the "superego" which gives normative value to certain concepts that guide what is considered "reality" within a society is a dynamic advance on Scheler's ideas. While Scheler understands that the behavior that attends "Ressentiment" may function from earlier memories that are not recognized as the source of negative behavior toward others, Freud is quite explicit about the dynamics between "unconscious" memories. The driving power of them through what he calls the "id," that is the emotional well of conative impulses that drive immediate responses, and the "ego" whose reflective, and thus, self-adjustive capabilities can with the rigor of Freud's "psychoanalysis" be enhanced. Such enhancement may with educative curricula, which are not yet a function of our public education, could create new normative understandings.

Freud raises individual psychology to its implications for communities, societies and nations. He does this not as an analog, rather as an extended process that began with origin of the small families that generated the human species. He takes a position, even as one who is essentially an atheist, and at best an agnostic, that the human race was meant to be a unity insofar as its intercourse with one another. The pathologies of social life and larger societies he will explain in this text, he feels, with an understanding of their origins, can lead not only to healthier single societies, but a world society:

> Just as a planet revolves around a central body as well as rotating on its own axis, so the human individual takes part in the course of development of mankind at the same time he pursues his own path in life. But to our dull eyes the play of forces in the heavens seems fixed in a never-changing order; in the field of organic life we can still see how the forces contend with one another, and the effects of the conflict are continually changing. So, also, the two urges, the one towards personal happiness and the other towards a union with other human beings must struggle with each other in every individual; and so, also, the two processes of individual and of cultural development must stand in hostile opposition to each other and mutually dispute the ground. But this struggle between the individual and society is not a derivative of the contradiction—probably an irreconcilable one—between the primal instincts of Eros and death. It is a dispute within the economics of the libido, comparable to the contest concerning the distribution of the libido between ego and objects. And it does admit of an eventual accommodation in the individual, as, it may be hoped, it will also do in the future of civilization, however much that civilization may oppress the life of the individual to-day.

---

5. I have not found any references in Freud's published works; for references from the 1870s, see Ralph Dumain, "The Autodidact Project: Quotes from Freud on Brentano & Other Philosophers." http://www.autodidactproject.org/quote/freud_brentano.html

The analogy between the process of civilization and the path of individual development may be extended in an important respect. It can be asserted that the community, too, evolves a super-ego under whose influence cultural development proceeds. It would be a tempting task for anyone who has a knowledge of human civilizations to follow out this analogy in detail. I will confine myself to bringing forward a few striking points. The super-ego of an epoch of civilization has an origin similar to that of an individual. It is based on the impression left behind by the personalities of great leaders, men of overwhelming force of mind o men in whom one of the human impulsions has found its strongest and purest, and therefore often its most one-sided expression. In many instances the analogy goes still further, in that during their lifetime these figures were, often enough, even if not always, mocked and maltreated by others and even dispatched in a cruel fashion. In the same way, indeed, the primal father did not attain divinity until long after head met his death by violence. The most arresting example of this fateful conjunction is to be seen in the figure of Jesus Christ, if, indeed, that figure is not a part of mythology, which called it into being from an obscure memory of that primal event. Another point of agreement between the cultural and the individual super-ego is that the former, just like the latter, sets up strict ideal demands, disobedience to which is visited with "fear of conscience" [p. 75]. Here, indeed, we come across the remarkable circumstance that the mental processes concerned are actually more familiar to us and more accessible to consciousness as they are seen in the group than they can be in the individual man. In him, when tension arises, it is only the aggressiveness of the super-ego which, in the form of reproaches, makes itself noisily heard; its actual demands often remain unconscious in the background. If we bring them to conscious knowledge, we find that they coincide with the precepts of the prevailing super-ego. At this point the two processes, that of the cultural development of the group and that of the cultural development of the individual, are, as it were, always interlocked. For that reason some of the manifestations and properties of the super-ego can be more easily detected in its behavior in the cultural community than in the separate individual.

The cultural super-ego has developed its ideals and set up its demands. Among the latter, those which deal with relations of human beings to one another are comprised under the heading of ethics.[6]

Freud in the above passages creates more than an analog between the individual and the society insofar as the dynamics of how the "libido," which can also be seen as a conative function of the "id" generates a positive or negative relation to others more or less spontaneously.[7] His ideas are a theory. There is a similarity to his argument that would eventuate in a world government, one wise in the ways of human judgment, to Kant's understanding of the need compromise or integration of self and others, in his essay *Universal History from a Cosmopolitan Point-of-View* (1784). In that essay, Kant saw the tension between the individual wanting to possess his or her own insights and discoveries for personal happiness, and the need to share them and be understood by others. There was

6. Sigmund Freud, *Civilization and Its Discontents*, trans. James Strachey (New York: W.W. Norton and Company, 1961), 88–89.
7. See Sigmund Freud, *The Ego and the Id*, trans. Joan Riviere (New York: W.W. Norton and Company, 1960), especially 45–47.

parallel work in the post-World War I culture in the social sciences to actually teach individuals the skills to listen to how they themselves interacted with others, and how to form effective cooperative relations. The field of group dynamics, which was the beginnings of social scientific praxis to facilitate interdependent understanding among groups of people, began in the 1920s, with an Austrian psychiatrist Jacob Moreno. He was the initiator of sociometry, psychodrama and group psychotherapy. His legacy has an essential thread: the primacy of relationships. His philosophy and strategies of research and therapy are as relevant today as they were then.[8] Soon, by the 1930s, theoretical insights into how to train individuals to be responsive to problems in personal relations to others, and how to facilitate cooperative groups will begin. This will be addressed below under the social scientific work of Kurt Lewin.

Freud in the above paragraphs brings up pathologies of the super-ego, and accurately describes the experience of Hitler with his father, knowingly or unknowingly. Moreover, Hitler's example is one of the misdirection of his "Ressentiment" upon the Jews, as in his life Jews were always personally helpful. The doctor who attended his mother's last days, Eduard Bloch, was appreciated and spared by Hitler from the extermination camps,[9] and, Hitler's recommendation for an Iron Cross, first class, came from a Jewish captain Hugo Gutmann.[10] Indeed, Hitler made a living with the help of a Jewish man who sold the postcards Hitler painted by the name of Neumann. Neumann even gave him an overcoat to wear as he did not have sufficient money to buy a new one.[11] The misdirection of Hitler's resentment and revenge was from a Freudian standpoint the unconscious reaction to his father's treatment of him.[12] The "ethics" of the Nazi party were those of punishing harshly those who were not Aryan, or, who disobeyed any of the seemingly arbitrary rules that emerged in Hitler's years of power.

B. Literature/Drama

Jean-Paul Sartre (1905–1980)

Sartre's play, *No Exit* (1944) has provided a close study of how past experience among individuals and the self-blindness that comes from what Scheler called "Ressentiment," and Freud "a pathology of the super-ego," can create as Sartre states at the end of the play "Hell is—other people."[13] Sartre was a Marxist, and a keen observer of how individuals interacted. One must appreciate that Marx from the onset of his thought saw interpersonal behavior as essential in comprehending how certain political-social norms

8. https://davidmalocco.wordpress.com/2015/05/08/jacob-moreno-1899-1974-psychiatrist/

9. https://en.wikipedia.org/wiki/Eduard_Bloch

10. https://en.wikipedia.org/wiki/Hugo_Gutmann

11. http://marcuse.faculty.history.ucsb.edu/projects/hitler/sources/30s/394newrep/394NewRepHanischHitlersBuddy.htm

12. https://www.theguardian.com/world/2005/aug/04/research.secondworldwar

13. Jean-Paul Sartre, *No Exit* in *No Exit and Three Other Plays* (New York: Random House, 1955), 47.

either facilitated or retarded interpersonal as well as personal understanding. Marx writes in 1844:

> Assume man to be man and his relationship to the world to be a human one: then you can exchange love only for love, trust for trust, etc. If you want to enjoy art, you must be an artistically cultivated person; if you want to exercise influence over other people, you must be a person with a stimulating and encouraging effect on other people. Every one of your relations to man and to nature must be a specific expression, corresponding to the object of your will, of your real individual life. If you love without evoking love in return—that is, if your loving as loving does not produce reciprocal love; if through a living expression of yourself as a loving person you do not make yourself a beloved one, then your love is impotent—a misfortune.[14]

The devolution of bourgeois authority means a redistribution of power among all persons who constitute a society. We all seek power in our environ. Sigmund Freud feels this drive to be recognized and given the power to shape ourselves, as well as others, is the core drive of human existence. Socialist society comes from this same need of which the bourgeois society has been formed, but unlike the uneven dialogues among the bourgeois insofar as cooperative values and action, the socialist inclined individuals recognize the central of mutuality and cooperation. Power can and must be shared if the separate integrities which are each person within a society are to be realized. Conversation, listening and modifying one's senses of exclusive authority is the manner of arriving at such equality. This understanding can be discerned and furthered through phenomenological reflection upon immediate conversation into the interpersonal dialogues of family, work and every other human activity among others. Karl Marx championed what will become an even more acute listening of how we converse, a phenomenological understanding of interpersonal dialogue, that is, he attended what was said between persons in a conversation, and how that language communicated thought, feeling and intention. Sartre's *No Exit* is just such a study.

The play is but one act, and in it we see no resolution or self-correction, rather styles of interpersonal thought and conversation that the "devil" has seen as creating the hell of being not listened to or not listening to self or others. For Sartre, this is a bourgeois hell, a hell of egotistical and pathologically self-deluded persons who cannot go beyond their own personal concerns or see sufficiently to self-heal. The play begins with one of three characters who will create Hell for each other. Garcin has just been led into the room in which he will spend eternity by the valet of the devil.

Garcin:	Hm! So here we are?
Valet:	Yes, Mr. Garcin.
Garcin:	And this is what it looks like?
Valet:	Yes.

---

14. Karl Marx, *Economic-Philosophic Manuscripts of 1844* in *Karl Marx, Friedrich Engels, Collected Works*, Vol. 3 (New York: International Publishers, 1975), 326.

Garcin: Second Empire furniture, I observe … Well, well, I dare say one gets used to it in time.

Valet: Some do. Some don't.

Garcin: Are all the other rooms like this one?

Valet: How could they be? We cater for all sorts: Chinamen ad Indians, for instance. What use would they have for a Second Empire chair?

Garcin: And what do you suppose *I* have for one? Do you know who I was? … Oh, well, it's no great matter. And, to tell the truth, I had quite a habit of living among furniture that I didn't relish, and in false positions. I'd even come to like it. A false position in a Louis-Philippe dining-room—you know the style?—well, that had its points, you know. Bogus in bogus, so to speak. (Sartre 1955, 3)

Sartre has Garcin situated in a chair, for eternity, in a room of the Second Empire style. Karl Marx had written a book on this time, *The Eighteenth Brumaire of Louis Napoleon* (1852), which Marx saw as "tragedy become farce." His meaning was that whereas Napoleon brought tragedy to France, his nephew Louis Napoleon brought only the farcical appearance of social equality. Sartre will characterize Garcin as a "farce" in his behavior, one who postures courage, but dies for a cowardly desertion in the military. He was a journalist, which symbolically, like Thomas Mann's reference to an artist as characterized by Plato, lives second-hand. His egotism has a false foundation, as he never encounters life full tilt. Why does that bring him to Hell. Here Sartre does not directly tell us as the three individuals speak of their sins. Garcin believes it was his abominable treatment of his wife for five years. That may be, but it is not the "sin" which brings him to Sartre's hell. In Sartre's eyes, simply being an untested egotist, one who has never risen to an act that corresponded to his self-boasting, is a sin. Louis Napoleon was infamous for his cowardice as he attempted to match his famous uncle in battle. Sartre, of course, participated in the French resistance early in the German occupation, but decided that such action was not him, and he turned to writing. Sartre's existentialism can be a template for his life in public. He always sought always to fulfill the projects to which he committed himself. Yet, could Garcin in some ways reflect his own self-doubt in this period, as in the play one cannot see the kind of "sin" in Garcin that is comparable to what one sees in the two others individuals who share the room with him for eternity, Inez and Estelle.

Inez is a lesbian woman, not a "sin" certainly in Sartre's eyes, but the characterization will play a role in how the three torture each other. Estelle is a woman who is essentially a narcissist. Inez lives to hurt others, as she characterizes it. Sartre more than likely understood her through the eyes of either Max Scheler or Sigmund Freud. Sartre studied phenomenological philosophy at the University of Freiburg in the 1930s, and was introduced to the ideas of a non-conscious consciousness in Husserl and Martin Heidegger. He depicts Inez as someone whose deep resentment, deep hurt is compensated for by her sadism. Inez put this into words:

Inez: When I say I am cruel, I mean I can't get on without making people suffer. Like a live coal. A live coal in others' hearts. When

I'm alone I flicker out. For six months I flamed away in her heart, till there was nothing left but a cinder. On night she got up and turned on the gas while I was asleep. Then she crept back into bed. So now you know. (Ibid., 1955, 27)

Estelle depicts herself and her inability to see others as individual persons in most of her dialogues. She is in Hell because she deliberately drowned her baby, and, the father, Roger, not her husband who she hid all this from, killed himself over her. Her conscience is not troubled. Her narcissism, however, is a contributing factor to her obliviousness to the harm she causes. Her self-characterization is untroubled:

Estelle: Excuse me, have you a glass? [Garcin does not answer.] Any sort of glass, a pocket-mirror will do. [Garcin remains silent.] Even if you won't speak to me, you might lend me a glass. [His head still buried in his hands, Garcin ignores her.]

Inez [eagerly]: Don't' worry. I've a glass in my bag. '[She opens her bag. *Angrily*] It's gone! They must have taken it from me at the entrance.

Estelle: How tiresome! [A short silence Estelle shuts her eyes and sways, as if about to faint. Inez runs forward and holds her up.]

Inez: What's the matter?

Estelle: I feel so queer. [She pats herself.] Don't you ever get taken that way? When I can't see myself I begin to wonder if I really and truly exist. I pat myself just to make sure, but it doesn't help much.

Inez: You're lucky. I'm always conscious of myself—in my mind. Painfully conscious.

Estelle: Ah yes, in your mind. But everything that goes on in one's head is vague, isn't it? It makes one want to sleep. [She is silent for a while.] I've six big mirrors in my bedroom. There they are. I can see them. But they can't see me. They're reflecting the carpet, the settee, the window—but how empty it is, a glass in which I'm absent! When I talked to people I always made sure there was one nearby in which I could see myself. I watched myself talking. And somehow it kept me alert, seeing myself as the others saw me … Oh dear! My lipstick! I'm sure I've put it on all crooked. No, I can't do without a looking-glass for ever and ever, I simply can't.

Inez: Suppose I try to be your glass? Come and pay me a visit, dear. Here's a place for you on my sofa.

Estelle: But—[Points to Garcin.]

Inez: Oh, he doesn't count.

Estelle: But we're going to hurt each other. You said it yourself. (Ibid., 1955, 19–20)

The torture of each other plays into each other's blindnesses. Estelle will not allow Inez's eyes to be her mirror. This deprives Inez of a heart to invest with a burning coal of unfulfilled desire. Garcin will at first ignore Estelle. Then, he will realize he needs someone to believe in him, believe in his courage. The exchange between Garcin and Estelle will typify eternity for him, as Inez has no interest in him, other than to insult him when he reaches out:

Garcin:	I want you to do me a service. No, don't shrink away. I know it must seem strange to you, having someone ask you for help; you're not used to that. But if you'll make the effort, if you'll only *will* it hard enough, I dare say we can really love each other. Look at this way. A thousand of them are proclaiming I'm a coward; but what do numbers matter? If there's someone, just one person, to say quite positively I did no run away, that I'm not the sort who runs away, that I'm brave and decent and the rest of it—well, that one person's faith would save me. Will you have faith in me? Then I shall love you and cherish you for ever. Estelle—will you?
Estelle [laughing]:	Oh, you dear silly man, do you think I could love a coward?
Garcin:	But just now you said—
Estelle:	I was only teasing you. I like men, my dear, who're real men, with tough skin and strong hands. You haven't a coward's chin, or a coward's mouth, or a coward's voice, or a coward's hair. And it's for your mouth, your hair, your voice, I love you.
Garcin:	Do you mean this? *Really* mean it?
Estelle:	Shall I swear it?
Garcin:	Then I snap my fingers at them all, those below and those in here. Estelle, we shall climb out of hell. [Inez *gives a shrill laugh. He breaks off and stares at her.*] What's that?
Inez: [*still laughing*]:	But she doesn't mean a word of what she says. How can you be such a simpleton? "Estelle, am I a coward?" As if she cared a damn either way.
Estelle:	Inez, how dare you? [To Garcin] Don't listen to her. If you want me to have faith in you, you must begin by trusting me.
Inez:	That's right! That's right! Trust away! She wants a man—that far you can trust her—she wants a man's arm around her waist, a man's small, a man's eyes flowing with desire. And that's all she wants. She'd assure you were God Almighty if she thought it would give you pleasure.
Garcin:	Estelle, is this true? Answer me. Is it true?

Estelle:	What do you expect me to say? Don't you realize how maddening it is to have to answer questions one can't make head or tail of? [*She stamps her foot.*] You do make things difficult … Anyhow, I'd love you just the same, even if you were a coward. Isn't that enough? [*A short pause.*]
Garcin: [*to the two women*]:	You disgust me, both of you.

And so it goes round and round for eternity. Is there a solution to this redundant pathology of self used by others? Group dynamics was recognized by Sartre, with some reservation which will be touched upon. That Sartre turned his attention to this manner of instructive intercourse as the premise of a more intelligent, effective society was as indicated Karl Marx's concern from the very beginning in the mid-1840s.

Sartre turned to the question of accurately integrating the autonomous mind and choices of individuals within the conditioning of their societies again and again in differing philosophical writings that included historical biography. His several volumes on Gustave Flaubert[15] were a way for him to comprehend how one might conceive the detours, dead-ends and avenues forward in human interpersonal discourse. What was remarkable about these studies was Sartre's movement of comprehending Flaubert both in the general society, changing society at that, of his times, within the small community of his family, and he himself insofar as how his original ideation was affected by these diverse collective conditions. This avenue of comprehending the individual within the conditions of the general society, an essentially Marxist premise, but one of the fourth phase in which both he and the earlier Marx began to write, is succinctly argued by Sartre in his 1960 book, *Search for a Method*.

In *Search for a Method* Sartre's premise is that one cannot simply define an individual by their class-interest or deep social background. Rather, there is an interplay between individual singularity and autonomy of the will, and the normative conditioning that is inescapable would one interact with others—an inescapable set of channels of immediate experiential life. Writing of how his approach differs methodologically from what he considers "a frozen Marxism," which Marx himself would eschew in that it ignored individual to individual dialogue and praxis. For Sartre, living Marxism is a "heuristic." Its concepts are meant, even by Marx himself, as but helpmeets to analyze how these general arguments are formulated and normed at specific times, as they are held in the emerging, progressing or regressing intelligence of individuals with their own life force. These larger sociological norms—"classes" and their "interests"—must be continually measured within how they are discerned in the praxis of the individual at a given time in history.[16] For Sartre, the method he employed in his study of Flaubert over the decades

---

15. See Jean-Paul Sartre, *The Family Idiot, Gustave Flaubert, 1821–1857,* 5 Volumes, trans. Carol Cosman (Chicago: University of Chicago Press, 1981).
16. See Jean-Paul Sartre, *Search for a Method,* trans. Hazel E. Barnes (New York: Borzoi Books, 1967), especially 25–31.

of his life was to at first establish the norms of a time, then go deeply into the individual thoughts and praxis of the person, and then show how these norms either exemplified or were challenged by that individual life. This last step, following the in-depth examination of the "particularities" of a life, is called the totalizing step. Writing of the "frozen Marxism" that belies this heuristic method, with the example of Georg Lukàcs, Sartre tracks how a "heuristic" has become only a scholastic flatness in contemporary Marxist thought among many:

> The totalizing investigation has given way to a Scholasticism of the totality. The heuristic principle—"to search for the whole in its parts"—has become the terrorist practice[17] of "liquidating the particularity" It is not by chance that Lukàcs—Lukàcs who so often violates history—has found in 1956 the best definition of this frozen Marxism. (Sartre 1967, 28)

One of the pioneers of the social science of group dynamics was begun by a German Social Democrat, as a way to address the problem of "frozen Marxism," before Sartre wrote. Sartre inherited the issue, albeit giving it his own significant understanding.

C. Micro-Sociology

Kurt Lewin (1890–1947)

Lewin began his studies of the interactions within groups in the late 1920s and early 1930s. He was influenced by Jakob Moreno's work in the psychodrama.[18] Thus, as a contributor to the study of individual interdependence within small groups, he benefitted from the initial work of this fourth phase by its first originator. Lewin's work over the years was both theoretical and pragmatic. Lewin moved to England and then the United States in 1933 because of Hitler's coming to power. Lewin was enlisted by the US government during World War II to help change the eating habits of domestic cooking, instituting a diet of cheaper cuts of meat, so that good meat could be sent to the soldiers overseas:

> When focusing on how to better educate and persuade gatekeepers, the most notable studies compared discussions with lectures. In the initial study that sparked many others throughout the war, Lewin (1943) investigated two methods of learning by offering various groups of Iowa housewives pediatric information about the nutritional merits of incorporating an unusual additive into their infant formula. Some of these groups of housewives were informed using a discussion-decision method, whereas others were informed using a lecture method. It was found that the groups informed by the discussion-decision method were three times more likely to consider and adopt infant formula. Given the success of this study, Lewin (1951) then

---

17. Sartre places a footnote here that reads "At one time this intellectual terror corresponded to 'the physical liquidation' of certain people." That is argued by historians who track Lukàcs's actions in Hungary between 1946 and 1953 https://en.wikipedia.org/wiki/Gy%C3%B6rgy_Luk%C3%A1cs#Communist_leader

18. https://en.wikipedia.org/wiki/Kurt_Lewin#Early_life_and_education

went on to examine these two education methods in the context of organ meat adoption, in which the discussion-decision method tended to generate nearly five times the level of trial as the lecture method.[19]

This emphasis upon fostering the thought and autonomous decision-making in a learning session was of the new democratic metaparadigm. One must be given the opportunity to think-through and determine one's own course of behavior, yet cognizant of the presence and needs of others. This fourth phasal inquiry and practice was of the desire to generate a democratic republic from the ground up. It hearkened to efforts toward "council democracy" in the European countries from late in 1918 to 1919, when giving more authority to people at a local level to choose their leaders was rampant. This direction of democracy will become more of a practice across the West in the idea of plebiscitary decision-making in our current metaparadigm. When in 1947, he began the National Training Laboratories, the site of his ground-breaking methods of interpersonal understanding and cooperation, in the United States, its first mission was to solve the problems of racism in Connecticut (https://en.wikipedia.org/wiki/Kurt_Lewin#Early_life_and_education). Lewin carried this out by filming the meetings of parents in segregated black schools and the majority white schools, and showing the thoughts to each non-included group, and then integrating the two groups of parents who now were informed of each other's thoughts and needs.

Lewin's early theoretical work on shared "life space" is a micro-sociology of how to realize a cooperative living and acting in mutual environments. The life space of a person changes with each new situation in which a person can make a choice of action, be it to engage a person, place or thing, or not engage. The concept of "life space" is a brilliant new one, possibly derived from Edmund Husserl's earlier concept of "life world,"[20] except the Lewin concept enables a close measurement of what he understands as a "field of forces" that condition the individual's choice. Husserl's "life world" is a more ambiguous, larger psycho-social area, not one that can be graphed. Lewin, for example, shows the invisible force field within a bathtub where two children bathe. Each child's "life space" is affected by the other's movement, which Lewin instructs us to recognize, as we all have such force field invisible boundaries in our immediate space to act or not to act.[21] Lewin describes the power dynamic of interaction of the two children:

> The two sixteen year old boys are sitting in a bathtub, the one very lively, excited, and over active, the other quieter. The excited one (A) jumps around in the tub so much that the other

---

19. Brian Wansink, "Changing Eating Habits on the Home Front: Lost Lessons from World War II," in *Journal of Public Policy & Marketing* 21, no. 1 (Spring, 2002): 95.

20. See Edmund Husserl, *Ideas Pertaining to a Pure Phenomenology and to a Phenomenological Philosophy, Second Book*, trans. R. Rojcewicz and A. Schumer (Dordrecht: Kluwer Academic Publishers, 1970), 385ff. Husserl conceived the concept that generally engages the mind and behavior of a person in the world that they act within during World War I. Revisions extended until the late 1920s, but it was published in English and in German by 1930.

21. Kurt Lewin, *Principles of Topological Psychology*, trans. Frtiz Heider and Grace M. Heider (New York: McGraw-Hill Book Company, 1936), 42–43.

(B) feels cramped. Finally (B) draws a line across the middle of the tub and tells (A) to stay within his own region. Whereas in the beginning there was a single unarticulated region as possible movement for (A) with the result for the other child (B) the actual freedom of movement was very restricted, now there are adjacent but sharply separated regions of free movement for (A) and for (B). (Ibid., 1936, 42)

Here Lewin provides a figurative drawing, a logical calculus of how the whole, the tub, is divided into parts. This kind of calculus can enable persons in any enduring mutual situation, or mutual changing situations, to define the spaces in which they may or may not cooperate. This is how the logical calculus of Leibniz in the first phase of the first Modern metaparadigm has been re-focused for mutual activity three hundred years later.

There can be several forces in the immediate environ that affect the individual's "life space," some material and some psychological. Certainly, the material actions of each child contribute to how the "life space" of each is felt, and congruent with the material determination, the psychological affects must be seen. Problem solving requires the variables to be discerned, yet here is a greater understanding of the specific variables than heretofore possible. Lewin writes of the "life situation and momentary situation," synonyms for a momentary "life space":

By psychological situation can be understood either the general life situation or the momentary life situation or more specifically, the momentary situation.

A woman stands at the loom in a big noisy factory, next to the last in the eighth row. A thread is broken. She is about to stop the machine to see what happened. It is shortly before the lunch hour. She has accomplished very little during the morning. She is annoyed.

These are a few data of the momentary situation of this woman. About her life situation one can say:

She has been married for three years. For a year and a half, her husband has been unemployed. The two-year-old child has been seriously ill, but today he seems somewhat better. She and her husband have been quarreling more and more recently. They had a quarrel this morning. Her husband's parents have suggested that she send the child to them in the country. The woman is undecided what to do about it.

It is obvious how closely life situation and momentary situation are connected. In this case, the life situation may serve as a rather remote background of the momentary situation. Or it can be that the woman was thinking of her child while she worked, and in this way the life situation often became part of the immediate situation. But even when she was busy repairing her broken thread and no longer thought of the domestic situation, even then the life situation remained at least indirectly significant. It affected the state of the person and thereby the reactions within the momentary situation.

The woman sees the momentary environment, the rooms, the bed, the household routine, in a different light with each change in the life situation. Objects which were dear to her before the trouble with her husband might have become disagreeable, others more precious. The room in which a child is ill changes its character and changes it once more when the child recovers. Their past history thus plays a great part in determining the psychological import of things for the person. (Ibid., 1936, 22–23)

Lewin certainly brought this logical calculus to the further step in his development of group dynamics, but it was not an explicit part of his training of others in the dynamics of cooperative self- and other understanding. What he developed by his early death in 1947 was a seven-step, multi-session workshop in cooperative action. The stated goals of the T-Group, with its emphasis on phenomenological listening and the formulation of concepts in conjunction with others, is the beginning of a post-group cooperation in the communities either with each other or the communities to which its participants return:

> ... the desired direction of learning and change is toward a more integrative and adaptive interconnection of values, concepts, feelings, perceptions, strategies, and skills.
>
> The advancement of such integration requires the learner, during any day of learning activity, shift in a more or less planned way among the roles of observer, diagnostician, evaluator, actor, and inquirer. In each role he deals with personal feelings of resistance, anxiety, threat, weakness, strength, euphoria, or satisfaction. Where inner and outer forces are discrepant and dissonant, the learner seeks to make changes in himself or in his environment which will establish greater congruence (with others). A similar cycle of vision and revision, inner and outer, is involved when a group or team is seeking more adaptive and creative ways of dealing with the environment of the larger social system in which it operates. [22]

The T-Group participants learn to be group observers themselves so that they can instruct their fellow-members as the group activity evolves. Moreover, there is a goal to enable the participant to bring these learnings to his or her own sites of vocation subsequent to this small-group experience. Seven major sets of skills then become the focus of T-Group training (Bradford, Gibb and Benne 1964, 84):

Skill Area I:	Assessment by the change agent (who is each participant) of his personal motivations and his relationship to the "changee" (the person whom is instructed in more effective behavior toward cooperative activity).
Skill Area II:	Helping "changee" become aware of a need for change and for the diagnostic process.
Skill Area III:	Diagnosis by change agent and change, in collaboration of their situation in terms of behavior, understanding and feelings to be modified.
Skill Area IV:	Deciding upon the problem, involving others in the decision, planning action and practicing the plans.
Skill Area V:	Carrying out the plan, successfully and productively.
Skill Area VI:	Evaluation as assessment of joint progress—methods of working and thinking and human relations.
Skill Area VII:	Continuing, spreading and maintaining accomplished changes.

---

22. Leland B. Bradford, Jack R. Gibb and Kenneth D. Benne, *T-Group Theory and Laboratory Method* (New York: John Wiley & Sons, 1964), 18.

Thereby, the T-Group approach in its "change agents" who return to their everyday vocations and other relationships is to facilitate through their own pedagogical interventions the gradual development of a society that moves toward solidarity in the many levels of organization that intersect from the ground up, i.e., "self-regulation and renewal from below."

Kurt Lewin's methodology when taken on by his successors, he having died in 1947, lacked the social change function for many reasons that were embedded in American norms of individual agency and existing societal structures. The most vocal and pointed criticism of the failings of the T-Group to fulfill Marx's vision of social change from the ground up as a result of the dialectical dialogues among citizen groups was voiced by Sartre.

It is worthwhile to quote Sartre at length in his attack on Lewin:

> In the work of Lewin, for example, (as with all Gestaltists), there is a fetishism of totaliza-tion; instead of seeing in it the real movement of History, Lewin hypostasizes it and *realizes* it in *already made* totalities. He writes: "It is necessary to consider the situation, with all its social and cultural implications, *as a dynamic, concrete whole.*" Or again: "The structural prop-erties of a dynamic totality are not the same as those of its parts." On the one hand, we are presented with a synthesis of externality, and to this given totality the sociologist himself remains external. He wants to hold on to the benefits of teleology while at the same time maintaining the attitude of *positivism*—that is, while suppressing or disguising the ends of human activity. At this point sociology is posited for itself and is opposed to Marxism, not by affirming the provisional autonomy of its method—which would, on the contrary, provide the means for integrating it—but by affirming the radical autonomy of its object. First, it is an *ontological autonomy.* No matter what precaution one takes, one cannot prevent the group, thus conceived, from being a substantial unity—*even* and *especially* if, out of a desire for empir-icism, one defines its existence by its simple function. Second, it is a *methodological autonomy.* In place of the movement of dialectical totalization, one substitutes actual totalities. This step naturally implies a refusal of dialectic and of history exactly because dialectic is at the start only the real movement of a unity in process of being made and not the study, not even the "functional" and "dynamic" study, of a unity already made. For Lewin, every law is a structural law and expresses a function or a functional relation between the parts of a whole. Precisely for this reason, he deliberately confines himself to the study of what Lefebvre called "horizontal complexity." He studies neither the history of the individual (psychoanalysis) nor that of the group. [23]

Sartre here attacks Lewin's "structuralism" where specific skills of listening to another or coordinating behavior with him or her can be improved as a distinct "part" in the service of the "whole" of cooperative *praxis.* This is "horizontal complexity" where skills exist in isolation from their historical cultures, rather as human potentials in any culture. Thus, for Lewin—in the eyes of Sartre—the skills and the outcome of their coordination are an

23. Jean Paul Sartre, *Search for a Method*, trans. Hazel E. Barnes (New York: Alfred A. Knopf, 1967), 58–59.

"ontological autonomy" that becomes his "methodological autonomy." Sartre addresses this as a false *praxis* inherent in the T-Group in his *Critique of Dialectical Reason*:

> And as long as the group, with its control mechanisms, remains effective and active, its fundamental truth is still *praxis*. However, we must retain this first aspect of *process* if we simply wish to mark the concrete limits of *praxis*. As long as one isolates it from the world in order to study it in its abstract purity, it will yield its intelligibility without transparency as an individual and common practice. When it is considered *in the world* without relation to anything except time and place, it will display new aspects: separations, scleroses, useless survivals, local wear, stratifications, the inertial force of apparatuses, fragmentation of the group, tendencies, antagonisms of function (carefully defined competencies cease to be so as *praxis* develops, because of the need to adapt to new circumstances). And the negative *praxis* of mediating apparatuses which attempt to dissolve these callosities and knots is essentially in danger of being simply an ever *anterior* liquidation, a preparation for common action, a restoration of instrumental functions without any other positive connection with the *praxis* of the group in the common field.[24]

Lewin did not isolate these listening and behavioral skills from the world. He was a change agent in public, and the skills prepared one to act in collaboration with others having engaged others in this mutual learning process. American culture eschewed active, combative change in the 1950s, and his successors, indeed, neglected historical issues as part of the training process. There is no reason in my view that the learned skills of the seven sets cannot gradually clarify inherited norms.

---

24. Jean Paul Sartre, *Critique of Dialectical Reason*, trans. Alan Sheridan Smith (London: Verso, 1976), 252.

Part IV

The Fourth Modern Metaparadigm, c.1970–c.2060

Chapter Thirteen

THE FIRST PHASE: SEMINAL IDEATION, C.1960–1980: THE FOCUS UPON DEFINITION AND HYPOTHESIS

A. Philosophy of History

Hayden White (1928–2018)

I will track White in this first phase, as well as the second and third phases of this present metaparadigm. He is an excellent example of how thought shifts with the demands of each phase. In one of White's first publications, in the mid-1960s, he writes an essay "What is Living and What is Dead in Croce's Criticism of Vico." The essay becomes part of a book White co-edited with Giorgio Tagliacozzo, a product of an earlier International Symposium on the late seventeenth and early eighteenth century historian, Giambattista Vico.[1]

White begins his essay with an interesting juxtaposition of the late nineteenth and early twentieth century historian, Benedetto Croce (1866–1952) with the late seventeenth and early eighteenth century historian Giambattista Vico (1668–1744). To make this the basis of his essay, White had to consider the final and formal cause of both Vico's methodology and the final and formal cause of Croce's methodology for their respective times of history. That meant, to carefully comprehend the manner of historiographical agency in Croce, and, correspondingly, not only how Vico wrote historical theory, but how he understood the sources of previous centuries that he considered. This is the conduct of "metahistory" by White, as it is for Croce, and, as we will see for Vico, who could be called the first metahistorian of Modernism.

One of Croce's main efforts was to show "what is living and what is dead" in prior philosophical systems, including among others that of Vico (ibid., 1969, 379). The care in discerning what is still fecund in a thinker from a past century is an ideational skill that demands a command of the ideation of one's present. Again, a "metahistorical" set of judgments. Yet, as we continue, we will see the arbitrariness of such judgments, and, how Hayden White addresses the issue of arbitrariness. White aptly focuses upon Croce's

---

1. Hayden White, "What is Living and What is Dead in Croce's Criticism of Vico?," in *Giambattista Vico, An International Symposium,* eds. Giorgio Tagliacozzo and Hayden V. White (Baltimore: Johns Hopkins Press, 1969), 370–90.

distinction between a "theory of history," and a "philosophy of history," as the seminal problems of both Croce and any other who would claim to be a "metahistorian":

> Croce distinguished between 'theory of history" and "philosophy of history." The former, he argued, was concerned to establish he criteria by which historians give to their narratives an appropriate form, unity, and content; the latter sought to discover the presumed laws by which human actions necessarily assumed the forms they did in different times and places. A theory of history was permissible, but only if it proceeded by means of a logic of intuitions, not a logic of concepts, that is to say, only if it were understood that history operated within the confines of art. In fact, the only conceivable theory of history, Croce held, was aesthetics. "Inasmuch as it is a science of pure intuition, a science of the individual object of pure intuition, aesthetics constitutes a theory of historiography." It was possible, then, to "philosophize" about the way in which historians, unlike "pure" artists, distinguished among intuitions "between the factually real (*réel de fait*) and the ideally possible." But—and here was the crux of the matter for Croce at that time—any attempt to "establish historical laws" had to be sternly suppressed. The search for laws was a scientific enterprise; science dealt with "the universal, the necessary, and the essential." History, by contrast, dealt with the individual, the empirical, and the transitory ("that which appeared and disappeared in time and space"). (Ibid., 1969, 382)

White would integrate the "theory of history" as aesthetic meaning with "philosophy of history" as a science with universal laws. This accomplishment takes historical thought, as well as literary thought, to an augmented level of "to know how oneself and others know" that is the new standard of knowing in human judgment. The way that he does this is to follow the advice of the phenomenological cultural historian and philosopher of history, Wilhelm Dilthey when he counsels Edmund Husserl how to more accurately comprehend how individuals structure their individual thought at a non-conscious level. He writes:

> Whereas a fixed delimitation was not possible for lived experiences, this could be found for expressions and objectifications … This indirect procedure that uses expressions has to some extent been applied by Brentano and Husserl. (Wilhelm Dilthey, *Poetry and Experience*, 1985, 229)

What Dilthey refers to when he speaks of "expressions and objectifications" is the attention to the "stylistics" of sentential judgments. Dilthey had written extensively on the early nineteenth century German who more or less developed stylistics as a discipline.[2] The surfacing of structures of verbal patterns of semantics and syntax that repeat themselves in a thinker, and, the preference for certain styles of sentential semantics and

---

2. Wilhelm Dilthey, *Leben Schleiermachers*, Two Volumes (Göttingen: Vandenhoeck & Ruprecht de Gruyer, 1966). See also Friedrich Schleiermacher, *Hermeneutics and Criticism and Other Writings*, trans. and ed. Andrew Bowie (Cambridge: Cambridge University Press, 1998); and, Friedrich Schleiermacher, *Hermeneutics, The Handwritten Manuscripts*, ed. Heinz Kimmerle, trans. James Duke and Jack Forstman (Atlanta, Georgia: Scholars Press, 1977).

syntax of a period of time, became for Hayden White the solution. It will be fleshed out by him in a systematic architectonic by his work *The Historical Imagination in Nineteenth-Century Europe*[3] in the second phase of this present metaparadigm.

Contemporaneous to White's discovery of how the use of grammar could be evidence of distinct laws of thought that convey distinct meanings was the work of Noam Chomsky, that brought the study of "deep structural" formation of sentences that were non-consciously developed as standard linguistic understanding.[4]

Vico's law of civilization change, the law of the *ricorsi*—the "spiral return" of distinct phases of social relationships with corresponding political and cultural institutions retrace this course on a similar, though significantly "metamorphosed" plane of existence or level of self-consciousness, was the focus of White essay as it came to a close (ibid., 1969, 385). This "spiral return," of course, is the thesis of this present text. White's awareness of this possibility of cultural development of periods of time we will see as a living potential in his work on *The Historical Imagination in Nineteenth-Century Europe*. The only other historian who had spoken of "spiral returns" of distinct phase of thought from century to century was Heinrich Wölfflin in his study of certain recurring perspectives in painting from century to century.[5]

White evidently from his later work pondered Vico's premise of the "spiral return," while being non-committal in this essay. What is clear from the essay insofar as White being an example of the first phasal ideation of the fourth Modern metaparadigm is his operational definition of how one is of an era in one's thought, but that an earlier era can augment one's thought with its own seminal and enduring truth. And, that to comprehend any historical thinker, one must situate oneself within the ideation of that thinker's time, later interpreters of that thinker, as well as one's own time, even if that means a self-realization that one's own thought may be not wholly contained by the norms of one's time. This conclusion is inferred by White in his final paragraph of the essay:

What, then, is "living" and what is "dead" in Croce's assessment of Vico's achievement? The clue to the solution of this problem is provided by two of Croce's judgments, one of Vico and one on himself. Summarizing his analysis of Vico in the last chapter of *La filosofia di Giambattista Vico*, Croce said that in the end Vico "was neither more nor less than the nineteenth century in embryo." And a few months later, in response to Borgese's "D'Annunzian" criticism of this book, he wrote that "... the philosophy with which I interpret and criticize the thought of Vico, while in some respects my own, ... is, in the main, nothing other than the idealistic philosophy of the nineteenth century." To be sure, Croce to have rendered it more "realistic" and more "critical" of itself; but in the end he remained within its horizons. Ample as they were, these horizons did not adequately encompass the operations of the physical

---

3. Hayden White, *Metahistory: The Historical Imagination in Nineteenth Century Europe* (Baltimore: Johns Hopkins University Press, 1973).
4. See Noam Chomsky, *Aspects of a Theory of Syntax* (Cambridge, Mass.: M.I.T., 1965), especially 15–18.
5. See Heinrich Wölfflin, *Principles of Art History: The Problem of the Development of Style in Later Art*, trans. M.D. Hottinger (Mineola, N.Y.: Dover Publications, 1950), 229–35.

sciences or of those social sciences founded upon similar aims and methods. Consequently, Croce's criticism of Vico did not really meet the main thrust of Vico's "new science," the effort for which many of the major socio-scientific theorists of the nineteenth century honored him. (Ibid., 1969, 389)

Thus, a metahistorical conundrum is there to be addressed as one reflects upon how Croce thought, to what degree was he of his own time, to what degree beyond that time, and did he properly grasp Vico and Vico's ideational time, and why Vico was important to the nineteenth century. Croce (1866–1952) began his significant works in 1900. He was really not "of" the nineteenth century, rather working within the framework of the deeper understanding of consciousness of his peers in many fields. That is why Croce stressed "intuition" rather than scientific laws. His understanding can be traced to Wilhelm Dilthey's ground-breaking book on the Humanities, and why they are not of the same causal laws as the material realities studied by the physical sciences.[6] While White misses this ideational stanchion that he himself will respect in Croce, nonetheless White did grasp the problematic of being a metahistorian of ideas. By the 1970s White will have addressed and solved this problem with his own new historical scientific model.

B. Drama

Judith Malina (1926–2015)

Judith Malina wrote in her diary on August 19, 1947: "Several films at the Museum of Modern Art: *Die Drei Groschenoper* with its allegorical finale, As the beggars disperse in the shadows. It is the interpolator again, the Greek chorus in modern guise. The chorus is dear to the audience because it bridges the gap—being part of both the audience and the secret mystery of the play."[7] Malina spoke of the future Living Theatre in her 1947 diary multiple times between June 4 and August 7 in 1947. Her reflections on August 19 are the germ of the idea we finally see in the 1960s and 1970s. The actors will present an analog of ordinary, contemporary life, and the moiety of the actors will become a chorus that draws the audience into the play—the play being them as they are in society. This drawing the audience into the play as its meaning, felt by compelled participation (or often willing participation) could be called a return to the "mystery" plays of the late medieval and early modern church:

> While modern theatre undoubtedly finds many of its origins in medieval drama, the mystery plays, pageants and morality plays of the 15th and 16th centuries are actually a very different animal, with quite a different set of associations. Imagine not a fixed stage and a darkened room, but mobile theatre out on the streets. Some people are watching carefully, but others are chatting with their friends, or buying food and merchandise from nearby vendors—keeping only one eye on the stage. Instead of the latest big name from an HBO

6. See Wilhelm Dilthey, *Introduction to the Human Sciences: An Attempt to Lay a Foundation for the Study of Society and History*, trans. Ramon J. Betanzos (Detroit: Wane State University Press, 1979).

7. Judith Malina, *The Diaries of Judith Malina, 1947–1957* (New York: Grove Press, 1984), 8.

series, you may well recognise a number of the actors from your own daily life. And instead of a focus on the individual and human relationships, you're treated to scenes from the Bible, about Christianity and the history of salvation.[8]

Indeed, when Malina realized the germ of her thinking in 1947 in a developed theatrical performance by the 1960s. She speaks of the "gestation" since 1946 (The Diaries of Judith Malina 1947–1957, 1984, 73). What will it be: the actors will leave the stage to mingle with the audience. People will be confronted where they sit or stand or they themselves will step forward. The willing or unwilling audience in their beliefs, values, everyday thought, will realize that they are in that performance consonant with the analog being presented on the stage, and now among them, guiding the problematic presented to them. Group dynamics practice had begun in 1947 at the National Training Laboratories in Maine, following the precepts of Kurt Lewin. How can interpersonal dialogue lead to mutual understanding, or, at least, clarifying differences? Malina in this fourth phase, the last phase of the third Modern metaparadigm, had creativity in the sense of a search for mutual truth in the actual persons participating in a here and now structure to inform her. She did know of John Cage in the late 1940s, and, met him in the early 1950s. He was formulating a theater among the audience in those years.

The move from the proscenium to the floor of the audience for the focus of the drama in the Modern era was first done by Luigi Pirandello (1867–1936) in his *Six Characters in Search of an Author* (1921). This was a stunning move, its purpose to dramatize that plight of working families in Italy, whose breadwinner often had to leave the region in order to support his family. Pirandello's values were of the new democratic world, but he became a supporter of Mussolini because of the syndicalism of his politics, that is, the emphasis upon the state supporting the maintenance of regional industry where families could remain in touch with their ancestral roots. Mussolini was a social-democrat, and then a national socialist. The staging of Pirandello had the same effectiveness in cultivating these democratic socialist values in audience that will be echoed in Judith Malina's "living theater."

Malina's *Mysteries* performed in the 1960s, was in the streets, confronting passers-by with a spectacle of nakedness and symbolic interactions among the troupe that allowed the ordinary people to join in. Here one can see how the "mystery plays" of the fifteenth and early sixteenth centuries were akin. The time of the "mystery plays" of the late medieval and pre- to early modern was a time of the conciliar movement, democratizing Christianity. It led to the individualism that became not only Lutheranism, but many other forms of such individual expression of the spirit, among these Zwinglianism and Presbyterianism. One can see a "spiral return" from this aesthetics of drama between the 1460s and 1560s and the emphasis of secular democracy upon the equality and sameness of the people in the problematics that bred democratic the institutions of the 1870s through the 1960s. Judith Malina, however, will create an even more radical individualism in her drawing out the voices of the individuals within her audience with their

---

8. https://www.bl.uk/medieval-literature/articles/medieval-drama-and-the-mystery-plays

elicited, but unexpected messages. This will heighten her problematic, for she and her fellow actors must know how to respond to dissidence as well as agreement. In a performance of the play *Paradise Now* at Yale University in New Haven, in September 1968, the actors not only engage the audience beyond the proscenium, the audience and actors in conflict or cooperation in rhythmic movements, chants, and political slogans, as well as other slogans of value-orientations, but leave the theater together and go into the streets.[9] Malina's diary relating the events of the street provide an insight into the radical experience, yet also its movement into the unresolved problem of the communication sought by the theater troupe:

> The last scene of *Paradise* moves out into the street. The night is warm and pleasant, and I think how nice it is that here there is not one to disturb, as the students are used to the sound of merriment in the streets. The students ride on the shoulders of the actors, the actresses ride on the shoulders of the students; the people move along York Street quietly and with the gentle elation that *Paradise* offers. I'm tired, and decide to return to the dressing room. En route I encounter Bob Brustein. He hands me a yellow flower, and with a consoling embrace he says, rather laughingly, "Judith, I must tell you, I *hate* this play." And we talked about that for a while. He feared that "this is a field left open to anything, even fascism." He fears that given freedom the people will choose fascism. Of course this is the fear that keeps the people from having their freedom.
>
> And as we are talking about this, a member of the audience comes to us, and bending down asks if I will sit up on his shoulders, and so I do, holding my yellow flower, and he walks with me toward the corner of York and Chapel. From my vantage point I see the police car ahead; as we come nearer I see two police with Windy between them.
>
> Windy is very young. A poet-adventurer, he has befriended the company. Everybody loves him.
>
> Windy's body is doubled down. I though he had been hit, because of the contraction of his body as it folded between the two cops. Later I learned they had used Mace. I said to the young man on whose shoulders I was riding, "Let me down, they're beating people up." As I was lowered to the ground, the students surrounding the police car were saying, "No violence, no violence." And the police heard them. And then the students sang "America."
>
> A cop pointed at Carl. I thought they might arrest him. I turned and said to the two people who had been walking with us that I was afraid there would be a bad scene; I started back to the theatre.
>
> As I reached the corner I saw Julian being escorted to the paddy wagon parked at the intersection at York and Chapel. There is a policeman on either side of Julian and another at the door of the wagon.
>
> As I approach I see Pierre Devis inside the wagon. Indicating Julian I say to the policeman:
>
> "That's my husband; can I go with him?"

---

9. See all but the last "scene" in the streets of a performance of *Paradise Now* in on video. Query online Judith Malina Living Theatre Paradise Now YouTube.

The policeman gestured to the wagon. I went inside, following Julian up the narrow steps.

Our two companions from the street had accompanied me to the wagon. "Get out of here," said the policemen to the bearded man with the colorful tunic. "I'll take my clothes off," said our friend. "Get in," said the cop.[10]

There are several problems that confront the intentions of Judith Malina and the Living Theatre as this final scene ensues. Her intention is to transform consciousness in this "here and now" of audience members as they enter the environment of their everyday life beyond the theater. However, this intention does not sufficiently recognize the laws of the city insofar as nudeness or a police officer's commands and judgment about disturbing the peace—even though this was not happening as described by Malina. Nonetheless, the law is the public. Malina realizes to some extent that she has fostered an understanding of anarchism that does conflict with one of her main messages in *Paradise Now*, "trust of others." The riding on each other's shoulders begins on the proscenium, extends down to the audience, engaging members of the audience in the earlier "scenes" of the play. Each scene is a differing form of troupe interaction, and, increasingly, audience interaction with the troupe. Falling from the proscenium into the hands of others is one of the scenes taking place. Sexual interaction is modeled, realistically, but with boundaries set insofar as a script. The limits are to exclude violence to one's chosen partner.

The problem of ending the play in the external environ is that it can't be controlled. In fact, some of the performances are unable to control violence within the theater (ibid., 1972, 47–48). That is why Bob Brustein's fear of fascism if anarchistic freedom allowed is justified. Malina, however, is seeking a level of knowing in a person that echoes Rousseau's counter to Hobbes—individuals, humans, are decent if they can win themselves back from the constraints of public law that bias certain behaviors and classes.

What we will see take place in the second phase of this present metaparadigm is a more controlled staging that is based on clear values the audience will understand. Thus, there will be a certain pull-back from the full spontaneity of taking the message with the troupe still engaged with you into the streets. The final act of the play *Seven Meditations on Political Sado-Masochism* (1973) will end with conversations in the audience between the performers and members of the audience. The seven meditations will be a successive series of values that form an architectonic of personal transformation—but one to be realized individually in one's own setting.

Malina was a writer and director always, and her metacognitive knowledge of theatrical history always enabled her to vary her performance according to a script, even as her script for *Paradise Now* had the kind of objective which was shared by the Encounter Group movement of the late 1960s—self-transformation "now.'

10. Judith Malina, *The Enormous Despair, The Diary of Judith Malina, August 1968 to April 1969* (New York: Random House, 1972), 42–44.

C. Group Dynamics: Advent of the Encounter Group

Morton A. Lieberman, Irvin D. Yalom (1931–Present) and Matthew B. Miles

In the 1973 text *Encounter Groups: First Facts,* Lieberman, Yalom and Miles offer case studies of 17 forms of group dynamic training that take place among students of Stanford University in the late 1960s.[11] The setting for the groups was a class on Race relations that lasted a semester, 15 weeks. They met once a week or some augmented times. Two hundred and ten students were enrolled in the 17 groups. Several of the training groups followed the traditional system of T-Group training, but by then what came to be called "Encounter Groups" sought a here and now focus on the individual, rather than seeking collective cooperation, were the majority of groups. Thirteen of the 17 groups were of either intensive encounter or two tape-led groups that offered a more low-keyed, self-chosen encounter with self and others. Even the traditional T-Group model incorporated some of the forms of self-revelation that will be a standard expectation for the Encounter Groups. Collective cooperation could be a goal, but the methodology was different from the active seeking of mutual understanding. Rather, as with the new historiography, and the Living Theatre, the individual was challenged to realize an overview of his or her resistances to listening and participating that compelled one to air to self and others aspects of personality that were not the material of T-Groups, which kept its focus on present behaviors.

In the concluding chapters of this text, the authors lay out a systematic architectonic of human functions that can constitute a plan for encounter groups that may follow, enabling more thoroughness, and an avoidance of "casualties," the negative affects suffered by many individuals who were challenged to a depth of mind and emotion of which they were unaccustomed. The always present danger and frequent actuality of casualties among the participants indicated the problem faced by the emphasis upon immediate recognition and change, a problem that certainly was present in the performances of the Living Theatre, and, in the new historiography, a more passive sense of avoidance among critical fellow-historians.

I will discuss the architectonic developed for Encounter Groups in the next chapter, which is the second phase of this present metaparadigm, when a systematic plan of inquiry and explanation guides the in-depth action-research of what will be the third phase.

Group #1 was a traditional T-Group. Yet, some of the methods were drawn from the "here and now" experience that challenged individuals in ways of which they were not accustomed. Leader #1 had 15 years' experience in T-Groups, which made him in 1969 one of the pioneer members of Kurt Lewin's training. But, as one will see in looking closely at any inquirer who practices through differing phases of the metaparadigm, new foci are inevitable given the time. Leader #1 uses a "trust" exercise that can be seen in the Living Theatre production of *Paradise Now* of 1968. "One member of the group at a

---

11. Morton A. Lieberman, Irvin D. Yalom and Matthew B. Miles, *Encounter Groups: First Facts* (New York: Basic Books, 1973).

time 'is cradled by others' and asked 'do you have any feelings that we might drop you?' "
(ibid., 1973, 23).

Group #3 is called by the authors a Gestalt group. The Gestalt groups did not use Gestalt therapy in the classic sense of its originator, Christian von Ehrenfels. Ehrenfels created a therapy in the third phase of the previous metaparadigm that used drawings or ink blots to elicit the unconscious associations of patients. Leader #3 formulated behavioral expressions that he modeled as venues for the group members. Perhaps that is why it used the Germanic term "Gestalt" or "form." The leader's forms of expression gave the members of the group a platform for self-revelation. The authors characterize the group in the words of its leader "love rest, letting it all hang out, pretend the pillow is her and let her have it, everything goes, fun, funny … (ibid., 1973, 28). Describing his leadership and activities, the authors write:

> One innovation of his has been the open sharing of his own fantasies and dreams. He reported that he pays careful attention to posture shifts, and helps members get in touch with themselves through these observations. He prefers to do group work because he finds it more exciting. He considers himself innovative and radical. He wants each member to have increased awareness about his own behavior and greater understanding of the "why" of such behaviors. He hopes that each member of the group will be able to choose to make his life what he wishes. He feels that it is important to frustrate members of the group in order to help them assume responsibility for their own acts. He tends to focus as much as possible on the here-and-now.

> Leader #3 was a very high support, high challenge, anti-intellectual leader who revealed himself freely, as observers saw him. The focus of his attention was fairly evenly distributed among interpersonal, intrapersonal, and group behaviors and midway among the leaders. He often invited questions and confronted members in an effort to "open them up." However, he gave a great deal of support (he and leader #8 were the highest). He offered friendship, as well as protection, to group members. With leader #4 he was the highest of leaders in his use of self. He revealed his here-and-now feelings and his own personal values, often drew attention to himself as a person and as a leader, and in many ways participated as a member of the group.

> He gave less attention than any other leader to any coherence-aiding statement, whether explanation, clarification, or interpretation. He managed the group by focusing on members or issues, or by suggesting some type of structural procedure.

> The observers rated his global style as "challenger," as "releaser of emotions by demonstration," and as a "personal leader," one who expressed considerable warmth, acceptance, genuineness, and caring for other human beings. (Ibid., 1973, 28–29)

In describing how the group was run structurally by leader #3, the authors write:

> Leader #3 used a high number of structured exercises, approximately sixty. Only two leaders used as many or more:

> Form small groups of four people. Take turns introducing yourselves non-verbally. Take five minutes to decide how you want to do it. Try to come up with a name for your group.

> Be a little metal box. Imagine you're empty: you're a shell. How do you feel?

I'd like some feedback. Tell me what you feel. Go around on me, on the leader.

Let's all try to bring you to a boil. I'd like to hear what the mind-fuck sounds like out loud.

Get in touch with your headache. Be your headache and talk to him. Say that's your head-ache. (Ibid., 1973, 30)

Leader #3 can be seen as iconographic for his time in the 1960s. As the same early years of Nietzsche in his first phasal philosophy of history, one's history surfaces in thought, but only to provide a cliff-edge from which to leap into the engagement of a sensuous life with others. History from this moment on is meant to be different. There are general definitions of what is being asked of one, and what kind of outcome is expected. But, as yet, no theories that can be systematically tracked in their personal outcomes, or the outcomes of others. One might compare Leader #3 to another first phasal mind, Henrik Ibsen, in his symbiosis of Brand—in his innovative daring, Peer Gynt—in his humor and life as enjoyment, and the final words of both plays that "caring" is he meaning of life.

Leader #8 made no effort to further the interpersonal dynamics of the group itself. In that, he can be seen as more of the nature of "encounter groups" as any, having completely departed from that tradition of group dynamics. He was seen as "charis-matic" by the observers (ibid., 1973, 51), an appellation that emerged in the mid- and late 1960s for every leader of cultural change. His methodology was called generally "transactional analysis," a school of thought in group growth that used intra-psychic work with one member at a time by the leader who was sympathetic to Carol Rogers's manner of asking a client to listen to what he or she said, and take ownership of what it represented—thus generating a self-analytic process.

Leader #8 used the "marathon" technique for group meetings. There were three weekly, three hour meetings, and then a 22 hour marathon over a weekend. The "mara-thon" was the non-stop meeting that could last a day and a night, used by many differing approaches other than "transactional analysis." All had in common the results looked for by the marathon, that is a loosening of defenses against criticism and self-criticism, so that in the spirit of John Locke, one could formulate a new ego position toward self and others. The kind of interventions by Leader #8 and the responses by members at the end of a marathon session are given to us by the authors:

Something's on your mind B ... Can I bring into the group questions you were asking me last night? I had the feeling that some of things you were asking had to do with your own dis-comfort. Do you want to talk about that at all? When you were talking about peer pressures, I thought you were talking about yourself and some of your own ideas. Is that accurate? Is there much pressure on you in your fraternity? ... Let me push my fantasy a little further. If it doesn't fit again, we'll drop it. My feeling was that you were in a very conservative college, and that you would not be one of the campus radicals ... I have a strong hunch you are afraid of disapproval. How do you feel about being talked about? ... Is that why you're afraid of arguing? ... Who hollers at you at home when you give your own opinion? ... What sort of things do you argue about? ... Are your parents narrow-minded? How are you going to try and change you? ... Hold it. D., what do you feel about this? (Leader asked everyone in the group about how they felt about what was happening.)

Would you do an experiment? … Would you get up and say to each person in the room some-thing positive about him or her? … Are you aware of how adept you were at picking out the strong points of everyone here? I think the important issue is that you don't have to look for funny things to say in order for people to listen to you. You are really in tune with what's going on. Who do you feel he's closest to? … You and J., want to sit in a room and talk about it for a few minutes? … What are you feeling now? … Any more things you want to say? … Anything else you want to do with B.? (This is extremely typical of the leader's behavior, asking each student whether or not he has gone far enough, or whether he wants to do anything else.)

You've got two choices about how to handle that. One is not to tell them about it and the other is to tell them about it. If you play along you feel badly, and if you don't play along you feel bad too … See you're playing "kick me". If you tell them and they respond, you feel bad … If you make decisions based on what is best for you, not upon child desires but upon what's important, then it's up to them to handle it … Well, that's the point. Do you want to go the rest of your life behaving in a way that will not hurt other people? It's actually their choice to feel hurt. They also have the choice of respecting you for making your own decisions … It sounds to me as if what's happening to you has put you in the position where no matter what you do, they're going to feel badly. If you don't tell them what you're doing, they'll feel badly doing what they don't want you to do … You've got to give that some thought, about their being hurt and your being hurt … Will you keep in mind that you are not in charge of their feelings? … They are in charge of their own feelings and you're in charge of yours. (Ibid., 1973, 51–52)

Leader #8 redirects all interpersonal responses to the individuals who make them. What is sought is more attention to the perspectives of each separate person, than to facilitate a cooperative group, as in the phase four of the previous metaparadigm in such group dynamic methods as T-Group Training.

Lieberman, Yalom and Miles, in the final chapters of this 1973 book, will provide a systematic architectonic of what essential values constitute how group dynamics should be viewed, and how to instill those values in group dynamic training. We will see an emphasis upon the intrapersonal, rather than the interpersonal. The interpersonal growth will await the fourth phase of the metaparadigm in which we are now. In the final chapter and final section of this text, I will touch upon some idea that may be fostered in the coming fourth phase of integration of tradition (which has become the T-Group method) and the new and challenging intrapersonal dialogues of trainer and individual member.

Chapter Fourteen

THE SECOND PHASE: DEVELOPING A SYSTEMATIC STRUCTURE FOR GUIDING NEW INQUIRY AND EXPLANATION C.1970–1990

A. Historiography

Hayden White (1928–2018)

Hayden White's *Metahistory: The Historical Imagination in Nineteenth Century Europe* creates a conceptual overview of four hierarchies of forms that have always been used in modern thought to structure the narrative of historical events. Each of the four hierarchies will consist of a foundational level of grammatical style, a narrative form of the emplotment of events furthered by the grammatical style, a well-known dramatic form reflectively honed that gives a cultural significance to the emplotments, and, an ideological implication, given the time of its political-social expression. He describes these forms in how they conjoin events with a thoroughness that enables a logical calculus of the differing semantics typical of four of the forms, that being their foundational sub-structure in what he calls their "deep structure," "pre-critical," or "pre-figural" expression. By that he means the non-conscious conscious level of the structure of thought discerned a century earlier by Brentano, Freud, Husserl and others who contributed to initial phases of deepening the functioning of human consciousness beyond the reflective level. All these four hierarchies in each of their four forms generate by their syntax and semantics at this deeper level of thought their spontaneous expression in the artifacts of writing or oral communication.

Having argued in his 1969 paper on Croce and Vico of the complex amalgam of normative and challenging structural logics of history that one must know, White now deepens that insight with a "stylistic analysis" that in some ways is superior to even that of Aristotle, found in his Rhetoric, and to some degree in his Poetics. White's 1973 text will offer a systematic guide into historical analysis of any age:

> This analysis of the deep structure of the historical imagination is preceded by a methodological Introduction. Here I try to set forth, explicitly and in a systematic way, the interpretative principles on which the work is based ... In this theory I treat the historical work as what it most manifestly is: a verbal structure in the form of a narrative prose discourse. Histories (and philosophies of history as well) combine a certain amount of "data," theoretical concepts for "explaining" these data, and a narrative structure for their presentation as an icon of sets of

events presumed to have occurred in times past. In addition, I maintain, they contain deep structural content which is generally poetic, and specifically linguistic, in nature, and which serves as the precritically accepted paradigm of what a distinctively "historical" explanation should be. This paradigm functions as the "metahistorical" element in all historical works that are more comprehensive in scope than a monograph or archival report.

The terminology I have used to characterize the different levels on which a historical account unfolds and to construct a typology of historiographical styles may prove mystifying. But I have tried first to identify the manifest—epistemological, aesthetic, and moral—dimensions of the historical work and then to penetrate to the deeper level on which these theoretical operations found their implicit, precritical sanctions. Unlike other analysts of historical writing, I do not consider the "metahistorical" understructure of the historical work to consist of the theoretical concepts explicitly used by the historian to give to his narratives the aspect of an "explanation." I believe that such concepts comprise the manifest level of the work inasmuch as they appear on the "surface" of the text and can usually be identified with relative ease. But I distinguish among three kinds of strategy that can be used by historians to gain different kinds of "explanatory affect." I call these different strategies explanation by formal argument, explanation by emplotment, and explanation by ideological implication. *Within* each of these different strategies I identify four possible modes of articulation by which the historian can gain an explanatory affect of a specific kind. For arguments there are the modes of Formism, Organicism, Mechanism, and Contextualism; for emplotments there are the archetypes of Romance, Comedy, Tragedy, and Satire; and for ideological implication there are the tactics of Anarchism, Conservatism, Radicalism, and Liberalism. A specific combination of modes comprises what I call the historiographical "style" of a particular historian or philosopher of history. I have sought to explicate this style in my studies of Michelet, Ranke, Tocqueville, and Burckhardt among the historians, and Hegel, Marx, Nietzsche, and Croce among the philosophers of history, of nineteenth century Europe.

In order to relate these different styles to one another as elements of a single tradition of historical thinking, I have been forced to postulate a deep level of consciousness in which a historical thinker chooses conceptual strategies by which to explain or represent his data. On this level, I believe, the historian performs an essentially *poetic* act, in which he *prefigures* the historical field and constitutes it as a domain upon which to bring to bear the specific theories he will use to explain "what was *really* happening" in it. This act of prefiguration may, in turn, take a number of forms, the types of which are characterizable by the linguistic modes in which they are cast. Following a tradition of interpretation as old as Aristotle, but more recently developed by Vico, modern linguists, and literary theorists, I call these types of prefiguration by the names of the four tropes of poetic language: Metaphor, Metonymy, Synechdoche, and Irony. Since this terminology will in all probability be alien to many of my readers, I have explained in the Introduction why I have used it and what I mean by the categories.

One of my principal aims, over and above that of identifying and interpreting the main forms of historical consciousness in nineteenth-century Europe, has been to establish the uniquely *poetic* elements in historiography and philosophy of history in whatever age they were practiced. It is often said that history is a mixture of science and art. But, while recent analytical philosophers have succeeded in clarifying the extent to which history may be regarded as a kind of science, very little attention has been given to its artistic components. Through the disclosure of the linguistic ground on which a given idea of history was constituted, I have attempted to establish the ineluctably poetic nature of historical work and to specify the prefigurative element in a historical account by which its theoretical concepts were tacitly understood.

Thus, I have postulated four principle modes of historical consciousness on the basis of the prefigurative (topological) strategy which informs each of them: Metaphor, Synecdoche, Metonymy, and Irony. Each of these modes of consciousness provides the basis for a distinctive linguistic protocol by which to prefigure the historical field and on the basis of which specific strategies of historical interpretation can be employed for "explaining" it. I contend that the recognized masters of nineteenth-century historical thinking can be understood, and that their relations to one another as participants in a common traditions of inquiry can be established, by their explication of the different tropological modes which underlie and inform their work. In short, it is my view that the dominant tropological mode and its attendant linguistic protocol comprise the irreducibly "metahistorical" basis of every historical work. And I maintain that this metahistorical element in the works of the master historians of the nineteenth century constitutes the "philosophies of history" which implicitly sustain their works and without which they could not have produced the kinds of works they did. (Ibid., 1973, ix–xi)

The thoroughness of the model presented reaches into what White calls the "deep structural" basis of judgment, and, its synonyms of the "pre-critical," and the "the pre-figurative." What White has in mind reaches back to the 1870s in Brentano's formulation of the non-conscious consciousness where sentences are "pre-figured," before reflective judgment. One can also understand this as Immanuel Kant's a priori judgments that emerge in the "productive judgment" (*Critique of Pure Reason*, 1965, 142–44 [A 118 – A 121]) which precedes reflective shaping of these linguistic forms that emerge spontaneously in our writing and speech. The "reflective judgment" is the shaping of this material into conceptual clarity for the reader (*Critique of Pure Reason*, 1965, 132–35 [A 101–106], White's higher level of historical narration. White has called himself an "unreconstructed" Kantian. In the Introduction to a book of Essays published in 1978, he challenges possible critics in his Introduction who might fault him for his Kantian epistemology:

> The essays in this book all, in one way or another, examine the problems of the relationships among description, analysis, and ethics in the human sciences. It will be immediately apparent that this division of the human faculties is Kantian. I will not apologize for this Kantian element in m thought, but I do not think that modern psychology, anthropology, or philosophy has improved upon it.[1]

In addition to knowing the historical norms of narrative in a past age, and, the narrative perspective of other historians, past and present, as well as one's own narrative preferences, knowledge of these deep structural elements also make one a "metahistorian." The discernment of a "deep structural" basis of judgment that conditions the facts that are

---

1. Hayden White, *Tropics of Discourse, Essays in Cultural Criticism* (Baltimore: Johns Hopkins University Press, 1978), 22.

narrated is a form of structuring thought that can be called "context free, insofar as it is present in all judgment, in all ages." Discernment of that reality, an aspect of the "spiral return" is the "meta" level of being a historiographer. White will go more deeply into this level which he calls the "tropological" ground of judgment in the essays of the third phase—the 1980s into his last writings before his death in 2018.

It is possible that White borrowed the term "deep structural" from Noam Chomsky, whose 1965 book, *Aspects of a Theory of Syntax*, spoke of sentence formulation in its complexity as "deep structure" movement, preceding conscious reflection, of syntax and semantics. White, then, has deconstructed and reconstructed narration to be of a new profound, multi-leveled consciousness. This was not his original idea—it began in the previous metaparadigm. What is new is his discernment of the several forms of historical judgment, and how they are carried by grammar.

White will augment his systematic architectonic later in the 1970s by suggesting how a logical calculus can indicate in the surface sentences the foundational level of semantic choices that generate emplotments, dramatic forms and the loaded language of political-social standpoints. He gave a lecture in 1974 at Yale University that offered the students, and when published in 1978, the public at-large this logical calculus to denote the syntactic and semantic choices that became the formal cause for the pre-figural levels that became the emplotments, the dramatic forms and the political-social rhetorical ideas:

Let us imagine that the problem of the historian is to make sense of a hypothetical *set* of events by arranging them in a *series* that is at once chronologically *and* syntactically structured, in the way any discourse from a sentence all the way up to a novel is structured. We can see immediately that the imperatives of chronological arrangement of the events constituting the set must exist in tension with the imperatives of the syntactical strategies alluded to, whether the latter are conceived as those of logic (the syllogism) or those of narrative (the plot structure).

Thus, we have a set of events

(1) a, b, c, d, e ...., n.
ordered chronologically but requiring description and characterization as elements of plot or argument by which to give them meaning. Now, the series can be emplotted in a number of different ways and thereby endowed with different meanings without violating the imperatives of the chronological arrangement at all. We may briefly characterize some of the emplotments in the following ways:

(2) $A, b, c, d, e,$ ....,n.
(3) $a, B, c, d, e,$ ....,n.
(4) $a, b, C, d, e,$ ...., n.
(5) $a, b, c, D, e.$ ...., n.

And so on.

The capital letters indicate the privileged status given to certain events or sets of events in the series by which they are endowed with explanatory force, either as causes explaining the structure of the whole series, or as symbols of the plot structure of the series considered as a story of a specific kind. We might say that any history which endows any putatively original event (*a*) with the decisive factor (*A*) in the structuration of the whole series of events following after it is "deterministic." The emplotments of the history of "society" by Rousseau, in his *Second Discourse*, Marx in his *Manifesto*, and Freud in his *Totem and Taboo* would fall into this category. So too, any history which endows the last event in the series (*e*), whether real or speculatively projected, with the force of full explanatory power (*E*) is of the type of eschatological or apocalyptical histories. St. Augustine's *City of God* and the various versions of the Joachite notion of the advent of a millennium, Hegel's *Philosophy of History*, and, in general, all Idealist histories are of this sort. (Ibid., 1978, 92–93)

White as he goes forward, however, will limit his logical calculus to semantics—key nouns and verbs that carry certain meanings further than one particular sentence. Probing the syntax of each sentential judgment could reveal the modes of emplotment that surface in the narrative, as well as the dramatic forms that emerge; but, it would require a choice by White. Subsequent to this essay, as he himself moves into what I term the third phase where his systematic architectonic is applied in depth in separate studies, he decides that he is talking more about the diachronic structure of an essay or book, and less about each sentence, each synchronic structure that constitute the whole sentential judgment by sentential judgment.

The examples given above by White are diachronic in scope. A synchronic logical calculus would take apart each sentence as characteristic of the individual historian's thought. Aristotle saw a distinct rhythm in a sentence from the onset of its grammar to its conclusion. He called this the "period" of the sentence (Aristotle, *Rhetoric*, 1409a–1409b [Book III, Chapter 9]). Moreover, Aristotle theorized about the affects on thought of differing syntax as well as semantics in a sentence: how antithesis in one sentence is created, the effects of metaphor, or of the longer metaphoric simile, of metonymy, of a stress on nouns of fact. Throughout Book III of his *Rhetoric*, he speaks of differing sentential styles of the "period" and its constituting grammar that he has discerned by his fellow Greeks.

White's avoidance of the single sentence for his logical calculus was influenced by Paul Ricoeur, of whom White wrote extensively. In an essay written in 1985 on Ricoeur, White indicates his eschewing of the single sentential judgment as a characteristic foundation for the whole in its narrative structure. This indication of the diachronic whole in preference over the synchronic sentence will explain the level of analysis never exacted by White. Even as the analysis of the semantics of sentences as they constitute in thought a trajectory of the whole as a form of emplotment and of dramatic style is in itself a ground-breaking new specificity for comprehending the "metahistorical" dimensions of the historical narrative. White writes in this instance of Ricoeur:

> For him, a narrative discourse is not analyzable into the local meanings of the sentences that make it up. A discourse is not, as some would have it, a sentence writ large; an analysis of a discourse carried out on the analogy of a grammatical or rhetorical explication of the

sentence will miss the larger structure of meaning, figurative or allegorical in nature, that the discourse as a whole produces.[2]

White will use his logical calculus to locate certain words that carry diachronic implications for the whole of one's argument in essay or book form. This is his focus upon the tropological terms that are conducive to certain kinds of emplotments. The tropological ground of poetic rhythm—metaphor, metonymy, synecdoche and irony—do function as characteristic in the construction of one sentence, creating distinct rhythms. But White will use the semantic implications and show them as the grounds of dramatic form—romance, tragedy, comedy, satire. All this is occurring at a "deep structural" level, which when surfaced, can be worked with reflectively.

What the tropes help constitute as well is the character of the complete diachronic argument. The argument is a whole which White, using the insight of a mid-twentieth century philosopher of aesthetics as well as rhetoric, Stephen C. Pepper, made central to his "metahistorical" understanding of the historical arguments possible, from age to age. These argumentative forms, Formism, Contextualism, Organicism and Mechanistic, were called by Pepper "the four world hypotheses."[3] Pepper conceived these organizational forms of judgment in the fourth phase of the third Modern metaparadigm, seeing them as differing perspectives that any individual could hold, and, that his teaching could enable a reader, a critic, a historian to find mutual cooperation by his or her recognition of these differing world views. Pepper was not an epistemologist, so did not theorize how these visions arose grammatically. He simply saw, quite clearly, these differences in the conceptual evidence of arguments. White speaks of them these "world hypotheses," giving full credit to Pepper, in over ten pages of his Introduction (ibid., 1973, 11–21).

For an Organicistic argument, one can see an employment of comedy—in the sense of all participants, as Peer Gynt, realize their humanity within their engagements whose interactions show the non-tragic nature of existence. For a Contextualist argument, one sees a satirical emplotment—in that no context is objective truth in itself, but is of cooperative or conflicting perspectives. For a Mechanistic argument one sees tragedy—in that there are winners and losers within the reality generated in thought.

At random, as it were, one can open White's book and see how these pre-reflective formative concepts emerge in any thinker. Writing of Hegel as an example of one who pursues a "mechanistic argument," with its "tragic" implications:

> The ideal state, Hegel noted, would be that in which the private interests of its citizens are in perfect harmony with the common interest, "when the one finds its gratification and realization in the other." But every actual state, precisely because it is a concrete mechanism, an actualization rather than merely a potentiality or a realization of the ideal state, fails to attain

---

2. Hayden White, "Ricoeur's Philosophy of History" in *The Content of the Form, Narrative Discourse and Historical Representation* (Baltimore: Johns Hopkins University Press, 1987), 172.
3. Stephen C. Pepper, *Word Hypotheses: A Study in Evidence* (Berkeley, Calif.: University of California Press, 1942; 1970).

this harmonious reconciliation of individual interests, desires, and needs with the common good. This failure of any given state to incarnate the ideal, however, is to be experienced as a cause for jubilation rather than despair, for it is precisely this *imbalance of private with public (or public with private) interests* which provides the space for the exercise of a specifically *human* freedom. If any given state were perfect, there would be no legitimate basis for that dissatisfaction which men feel with their received social and political endowments, justification for the moral indignation which stems from the disparity between what men desire for themselves and feel, because it is the only criterion of right they *immediately* feel, to be a *morally justifiable* desire, and what the community into which they are born are asked to live out their lives insists that they *should desire.* Human freedom, which is a specifically moral freedom, arises in the circumstance that no "present" is every adequately "adapted to the realization of aims which [men] hold to be right and just." There is always an unfavorable contrast between "things as they *are* and things as they *ought* to be." But this precondition of freedom is also a limitation on the exercise of it; every attempt to correct or improve the state, by reform or revolution, succeeds only in establishing some new mechanism which, however superior it may be to what came before, is similarly limited in its capacity to reconcile private interests and desires with the common good and needs. (Ibid., 1973, 108–9)

White's remark that one should feel "jubilation" over this Hegelian position of the failure of any given state because the imbalance of the private and the public, because it is the threshold of personal freedom, of personal autonomy, which is not a "tragic" interpretation as that of Hegel. Rather, it is an organicist interpretation whose emplotment is "comedy." Indeed, White's ability to discern many differing understandings of history in his array of historians is made possible, in part, by a "comedic" view of historical events. History for White is not tragic, rather it is the spiral return of all these differences and their trajectories in thought and action—an "organicist" whole, ever progressing in knowledge, but never finished.

What White does not make clear in his use of Pepper's analytical argument forms is that these are inherent in a thinker over that thinker's career of thought, which White argues in characterizing a thinker. This appreciation is significant in understanding what areas White understands as "deep structureal" and "pre-figurative" in his epistemology.

White has given historical thought in its broadest sense of non-professional as well as professional reflection upon the past a systematic architectonic that can be more finely-tuned. Yet, it is a step forward into the problem-formulation and problem solving that will take us all toward a greater chance at world cooperation through mutual understanding. [This author is a Formist-romanticist.]

B. Drama

Judith Malina (1926–2015)

Judith Malina became far more conceptual in her dramatic choreography in the search of fostering through her work a more self-conscious individual who is equally conscious of the other person's presence and thoughts. Being aware of how one is aware, and how the other is aware, as well as those conventions that keep us from this greater knowing, was more artfully sought in her 1973 drama *Seven Meditations on Political Sado-Masochism*.[4] The drama is in seven scenes, one for every conceptual self-understanding and understanding of the other person deemed necessary to revolutionize the world. Thus, spontaneity in response is not actively sought, but can occur during the six initial meditations. The seventh scene, "revolution," will be a greater contact with the audience, but in the form of rational conversations initiated by the actors. This rational development of a systematic inquiry with a focused response to each attendant concept will mirror all second phasal approaches in the arts and the sciences, guided by the systematic architectonic each of these disciplines develops to break new ground and flesh out in the third phase of conflict with tradition.

As late as 2012, in an interview that included a current member of the Living Theatre, we hear that *Seven Meditations on Political Sado-Masochism* was singular in its fostering of conceptual reflection. The young, contemporary actress, a member of the present Living Theatre cast, also speaks of her work as a singular performer who tells episodes of her life story, in depth, the third phase of the current metaparadigm, of which we are now in.[5]

The overarching concept of political sado-masochism is addressed with the following meditative scenes, in sequence:

(1) sexual love, and its repression and subversion through the normative conventions of the society;
(2) authority, and its contemporary societal expression;
(3) the master–slave relationship, a consequence of the present conventions of authority;
(4) money, as the medium of exchange, a lifeless form used for subjection in the master–slave relation, as well as its concomitant unequal distribution;
(5) violence by the policing authority of the society;
(6) death and darkness of the soul, the consequences of the above norms; and
(7) revolutionary change, through liberation and anarchy.

Malina took these principles from Leopold Sacher-Masoch, whose name was given to his own original research into societal pressures that generated social-sexual pathologies. Sacher-Masoch had created his own architectonic system of life in society in its negative aspects from 1870 to the early 1880s in a series of novellas, grouped under the title called *Das Vermächtnis Kains [The Legacy of Cain]*. Interestingly, the years in which the principles were developed into a sequential interrelated architectonic were of the second phase of the previous metaparadigm. Malina, who saw the need to draw back from the unchecked spontaneity of response from members of the audience, both in and without the theater, found Sado-Masoch's approach of offering the public clearly defined principles that constituted the core of societal alienation and its perversions fecund for her present. Malina gave them her own up-dated thought, adding, of course, the choreography and stagecraft to Sacher-Masoch's principles that made them more sensuously present as a focus for meditation.

Sacher-Masoch's six principles did not go further to a seventh, which was "revolutionary change, through liberation and anarchy." Below are his six principles each having one or more novellas whose plot briefly explains his perspective[6]: He ceased writing on these principles, as he had, in the mid-1880s.

(1) Love
      Prologue: The Wanderer
      1. Don Juan of Kolomea
      2. The Man Who Re-Enlisted (1869)
      3. Moonlight (1868)
      4. Venus in Furs (1869)
      5. Plato's Love (1870)
      6. Marcella (1870)
(2) Property
      1. People's Court
      2. Haidamaka (1877)
      3. Hasara Raba
      4. A Testament (1875?)
      5. Basil Hymen (1875?)
      6. The Paradise on the Dniester.
(3) The State
      1. Ilau
(4) War
(5) Work
      1. The Old Castellan (1882)
(6) Death
      1. Frau von Soldan (1882)
      2. The Jewish Raphael (1882)
      3. The Godmother (1883)
      Epilogue: The Night Before Christmas [not written]

Judith Malina's treatment of her take on Sacher-Masoch's six principles is designed as a choreographed set of meditations. Throughout all six there is a meditative hum with a rhythmic flow that never stops, even as the individual actors/dancers perform their enactments of the meaning of that meditation. The actors sit in a circle, at the head, he who is the chief authority throughout all six phases, with only the seventh changing as the actors liberate themselves through individual moves that include turning around and speaking with members of the audience. The audience sits in a circle around the actors through the performance.

---

6. https://en.wikipedia.org/wiki/Legacy_of_Cain

In each scene, including the seventh of liberation, one of the actors reads out a prepared set of axioms that guide the thought to be transmitted that guides the actions of the performers. In sexual love, we see the compulsive compliance by the passive victims' relationships, where the water of erotic life is dispensed by the authority and punishments of violation are given by a stick blow. In the scene of authority, chains have been introduced and willingly put on. In the master–slave relation, some members are masters and the others willing slaves, the former dragging the latter in chains across the floor. In the abstract medium of interpersonal exchange, money is counted as the means of communication, which leads to violence by the authorities. Most moving is the death and darkness of the soul scene, where all the individual members, even he who was the chief authority, moan and contort in a moving manner against the meditative hum.

This play will continue to be produced as an essential component of the Living Theatre's repertoire, even as new in-depth plays, conceptual as well, figure into the third phase of the 1980s into the present.

C. Group Dynamics: The Guiding Principles of Encounter Groups

Morton A. Lieberman, Irvin D. Yalom (1931–Present) and Matthew B. Miles

Writing on the "new technology" of group dynamics, the encounter group, in a later chapter of the 1973 work, *Encounter Groups: First Facts,* the authors state:

> If the chief effect of encounter groups is to heighten the valuation of personal change, it must be remembered that they have also been thought to have provided a new technology for human change and growth ... Four basic dimensions were found to describe the activities of encounter leaders: Stimulation, Caring, Meaning Attribution, and Executive Function. Except for Stimulation, these dimensions in a general way will remind those familiar with the small group literature of other "functional theories" of leader behavior in small groups. Certainly what we have termed Executive Function, the management aspects of leader behavior, and the Caring function correspond to the two parameters postulated by Bales,[7] the oft-cited socioemotional and task functions of leadership. Meaning Attribution, a tongue-twisting abstraction to describe an array of leadership functions that have to do with interpreting or labeling or other behavior aimed at creating a cognitive perspective in the participants is again a dimension of leader behavior that has been observed in studies of task, therapeutic, and educational settings. While the dimensions which emerged in studying encounter leader behavior are not identical to those described in earlier studies, they do share similarity with other analyses of leader behavior. Only the dimension of encounter leader behavior which we have called Stimulation seems to represent an unusual technological contribution of the encounter group movement. Certainly stimulation represents a core aspect of "inspirational" teaching or leadership. Yet, in the past decade educators and psychotherapists have increasingly eschewed excessive stimulation. Recall that Stimulation

---

7. See R.F. Bales (1953) "Task and accumulation of experience as factors in the interaction of small groups." *Sociometry* 16: 239–52.

encompasses a large variety of encounter leader behaviors such as confrontation, personal revelation, demonstration, stagings, which in the eyes of some have become synonymous with change-oriented groups which depart from earlier forms. Our findings speak clearly that Stimulation is not *in and of itself* a productive behavior and leaders very high on this characteristic were not only modest in gain but high producers of negative outcomes. (Ibid., 1973, 447–48)

The insight that "encounter" in itself is a negative technology occurs in 1973, the precise year in which Judith Malina retreats to a more conceptual approach of the Living Theatre with *Seven Meditations on Political Sado-Masochism* (and which a century before lead to the same movement in literature by Leopold Sacher-Masoch). This, of course, can be a year that is traced back century by century, even though second phasal systematic architectonics does not always occur so precisely. Hayden White's *Metahistory: The Historical Imagination in Nineteenth Century Europe* also shared that date.

Lieberman, Yalom and Miles call "encounter" as "stimulation," a technology. Why? They see it as a reflective choice. But, in the perspective of metahistorical ideation it is more. It is of the aggressive, challenging call we see in the first phasal message of the second Modern metaparadigm of Edward Young in his view of the genius, in Nietzsche's "Übermensch," as well as in the early 60s' Judith Malina. A technology is more than a systematic lever, it is also a way of thought and behavior. Lieberman, Yalom and Miles then ask, can training take the danger out of "encounter" as a technology?:

Could our conclusion about the relative impotence of encounter as a mechanism for personal change be erroneously explained as a function of the inadequacy of the technology rather than the competence of the leaders? We think not. Most of the leaders were prestigious, "senior" encounter leaders, highly experienced and well-regarded representatives of their orientations. In our own observations of these leaders as they worked with participants we tried to ask how well each leader applied his own framework of "theory of change". ...

What went wrong? As the use of encounter groups expanded at an explosive rate, and the prophets became many, the basic assumptions about what it is people need and under what conditions they can grow and develop were stretched into grotesque shapes with a loss in perspective and balance. Some encounter leaders distinguished themselves on the principle that if freeing up human beings to express emotions was good, then progress could be enhanced, intensified, and speeded up by greater and greater dosages. For others, if developing closeness and the basic human respect and trust for others was a good, then more and more love made faster and expressed at increasingly intense levels would be even better. For others, if the ability to talk about previously unrecognized or unspoken aspects about oneself was a good, then the total stripping of self and the total closure of all hidden recesses was even better, and techniques were developed to facilitate and speed up this process. If providing a setting for human growth and enhancement required that this setting not be constrained by the ordinary mores and mutual expectations of society, then a total and instantaneous sweeping away of the fabric of mutually acceptable restrains in social relationships was seen as the answer. The encounter group scene became like a science fiction portrayal of basically sound procedures; procedures so misshapen by excess that they no longer served their original function. (Ibid., 1973, 449, 454–55)

What these authors stress indirectly in the above discussion, and more explicitly in the last chapter throughout, is that the three other principles—Caring, Meaning Attribution and the Executive Function—must create an environment of learning that can adequately integrate into the existing norms of society. This pull-back is necessary, but not wholly, when looked at from the historical political-social perspective. Change agents do change the ideation of culture. Judith Malina's Living Theatre is an example of long-term efforts in social-political change. Nonetheless, Lieberman, Yalom and Miles do see how a balanced approach to even people "in transition" is necessary for long term, effective change.

Chapter Fifteen

THE THIRD PHASE: MATERIAL INQUIRY INTO THE VERIFIABILITY OF SPECIFIC CONCEPTS, AND CONFLICT OVER THE IMPLICATIONS OF THE FINDINGS C.1990–C. 2020

A. Philosophy of History

Hayden White (1928–2018)

White, writing in the mid-1980s, addresses specific topics in his systematic, hierarchical architectonic of historical thought. Essays that take up certain topics, such as that of individual historians, tangential to his work, or the problem of temporality in narrativity, narrativity as a story form,, enable him to probe deeply into how past and present issues were formulated, and in many instances seen as seminal to his present thought. Looking at one essay, published in 1987 by White, we can see how the third phase of conflict through material argument (evidence) and its implications are developed. In "The Question of Narrative in Contemporary Historical Theory," his views on narrative and temporality are presented, most deeply and clearly.[1] White argues that the "form" of the narrative is always a "story." Herodotus, of course, used his Greek term *historia* to mean both inquiry into the historical facts and a narrative "story" of what occurred. In this way, White's understanding, with the added dimension of how temporality is effected and how it affects thought, is that of Herodotus. The German language also has this bipolar meaning in the term "Geschichte," which is both "history" and "story."

White explains this conjunction of "story" and a chronology of factual occurrence beginning with a quote from Paul Ricoeur:

> "The plot … places us at the crossing point of temporality and narrativity; to be historical, an event must be more than a singular occurrence, a unique happening. It receives its definition from its contribution to the development of a plot."[2]

---

1. Hayden White, "The Question of Narrative in Contemporary Historical Theory," in *The Content of the Form, Narrative Discourse and Historical Representation* (Baltimore: The Johns Hopkins Press, 1987), 26–57.
2. Paul Ricoeur, "Narrative Time," in *The Philosophy of Paul Ricoeur: An Anthology of His Work,* ed. Charles E. Reagan and David Stewart (Boston: Beacon Press, 1978), 171.

According to this view, a specifically historical event is not one that can be inserted into a story wherever the writer wishes; it is rather a kind of event that can "contribute" to the "development of a plot." It is as if the plot were an entity in process of development prior to the occurrence of any given event, and any given event could be endowed with historicality only to the extent that it could be shown to contribute to this process. And, indeed, such seems to be the case, because for Ricoeur, historicality is a structural mode or level of temporality itself.

Time, it would appear, is possessed of three "degrees of organization"; "within-time-ness," "historicality," and "deep temporality." These are reflected, in turn, in three kinds of experiences or representations of time in consciousness: "ordinary representations of time, ... as that 'in' which events take place"; those in which "emphasis is placed on the weight of the past and, even more, ... the power of recovering the 'extension' between birth and death in the work of 'representation'" and, finally, those that seek to grasp "the plural unity of future, past, and present." [White 1978, 171] In the historical narrative—indeed, in any narrative, even the most humble—it is narrativity that "brings us back from within-time-ness to historicality, from 'reckoning with' time to 'recollecting' it." In short, "the narrative function provides a transition from within-time-ness to historicality," and it does this by revealing what must be called the "plot-like" nature of temporality itself. [Ibid., 178] ...

Thus, in telling a story, the historian necessarily reveals a lot. This plot "symbolizes" events by mediating between their status as existents "within time" and their status as indicators of the "historicality" in which these events participate. Since this historicality can only be indicated, never represented directly, the historical narrative, like all symbolic structures, "says something other than what it says and ... consequently, grasps me because it has in its meaning created a new meaning." [Ibid., 1987, 51][3]

For Ricoeur the "whole" of the story form, be a romance, a tragedy, a comedy or a satire imparts a meaning that is more than the facts or the sentences that carry the facts. "Deep temporality" is carried by the story form. While the form of story in its beginning, middle and end is a shape that uses "within time" for the facts, and, in its succession creates a "historicality," as one ponders what became before and after, it also imparts its own character of time in its "deep symbolic structure." The "deep structure" is how one's thought casts the meaning of the narrative as a narrator, and how the discerning reader is impacted by this story form. White's theory of emplotments is keyed by Ricoeur's idea of the "deep temporality" of story forms. White's epistemology, which credits Kant, Freud and others who see a pre-reflective consciousness forming expressions of meaning, even when the narrator is unaware of the form being created, can supplement Ricoeur's ideas with such concepts. Moreover White's idea of an even deeper pre-reflective consciousness that creates the temporalities of formism, contextualism, organicism and mechanism, enables him to see how individuals unschooled in the classic forms of story generate the grammatical syntax and semantics for such structures, which after all were invented by minds inclined to see reality in such "story" forms.

---

3. Paul Ricoeur, "The Language of Faith," in *The Philosophy of Paul Ricoeur: An Anthology of His Work*, ed. Charles E. Reagan and David Stewart (Boston: Beacon Press, 1978), 233.

White uses the epistemology of Kant, but in a way that doesn't dwell on the character of single sentences, as Aristotle did in his understanding of the "period" of rhythmic attention created by the sentence structure. Husserlians picked up Aristotle in this sense of seeing temporality in its meaning of "deep temporality" generated by characteristic judgments of single sentences.[4] David Carr, a Husserlian phenomenologist, attacks White's epistemology of "deep temporality" carried by the story form at length in his 1986 book *Time, Narrative, and History*.[5] Carr counters White by referring to the pre-reflective duration of judgmental attention in single actions (see especially, Carr 1991, 21–40 and 45–57). Each action has a rhythm, a duration, which is equivalent to the Aristotelian "period," although not necessarily of grammar. Grammar may be an element of it, but Carr dwells justifiably on "attention" and its relevance in generating temporality (see Kant, *Critique of Pure Reason*, 1965, 168, n. [B 157]). All such acts are an event or cumulatively are an event. The "deep temporality" for Carr, and indeed, for Husserl, is this aggregative sum, which can incidentally form a "story," but are not conditioned at the "deep structural "level by classic story forms. White's conception, taken from Stephen C. Pepper, of the deep structural forms of human attention and knowing in either the formist, organicist, contextualist or mechanist modes, undercut, Carr's criticism of White. It is not the deep structural imparting of a classic form of story that was reflectively created by poets and authors, rather the deep structural mode of attending reality—to date only perceived by a few scholars like Pepper and White—that has been recognized as the fundamental ground of knowing at the pre-reflective level.

What can be developed in the philosophy of history in the fourth phase of the metaparadigm in which we are now in, given this background of the new epistemologies of historical knowing? One possibility is to bring White's and Pepper's insights into how we analyze events to a level of a society's normative approach. As Freud did in speaking of pathological societies, there can be evidence of how a nation has a normative way of analyzing events that while not embedded in individual "deep structural" consciousness, has been derived from leading thinkers over the history of that culture. This would correspond to an effort to understand how collective thought in a culture might condition individual styles in ways not yet explored, as well as how individual styles are necessary for the collective normative thought of the same culture.[6]

---

4. Edmund Husserl, *Logical Investigations*, Two Volumes, trans. J.N. Findlay (London: Routledge & Kegan Paul, 1970), 2: 484–89 [Par. 25]. Husserl calls the several clauses and phrases that constitute the sentence "time-stretches." The entire sentence is called a "temporal concretum." One's attention is carried through the thought, a micro-history of how one attends. The "temporal concretum" would be the form that finds it cultural form as a diachronic image of the "story," the replication of that temporal concretum as a narrative form.
5. David Carr, *Time, Narrative, and History* (Bloomington: Indiana University Press, 1991), 8–9, 11–16, 19–20, 59, 72, 88–90 and 95.
6. See, for example, Mark E. Blum, *German and Austrian-German Historical Thought in the Modern Era* (New York: Lexington Books, 2020).

B. Literature

Elfriede Jelinek (1946–Present)

The new aspect of actor–audience interaction that we have seen in the Living Theatre, that is the effort to stimulate the audience to respond with a self-probing analog or argument in the here and now to what ideas are being articulated by the actors, is in literature the same in these initial stages of the metaparadigm. Both Jelinek and Sebald use a device in literature one can trace to Goethe's *Sufferings of Young Werther,* that is, the "unreliable narrator," but with augmentations that challenge the reader to perform a probing self-analysis in order to come to know how and why the narrator is unreliable.[7] This effort is prompted by narrative devices that can be found in the history of modern narration, one prominent being "erlebte Rede,"[8] where an objectively distant voice, while describing a protagonist suddenly in the sentence allows us to see an experiential thought or action that demands our own emotional reaction in order to comprehend it. Dorrit Cohn, a contemporary literary scholar, calls this "a mimesis of other minds" demanded of the reader (Ricoeur 1984, 91).

Elfriede Jelinek, **The Piano Teacher, A Novel**[9]

Jelinek uses "mimesis of other minds" as her novel begins and throughout the novel into its last sentence, trusting the discerning reader will have completed a self-analysis in the process. "Mimesis of other minds" is especially present in a marked way where the shift that the reader must attend may be in a single sentence, but also where a following sentence or several is at once an objective description of a thought or act, and simultaneously a window into the decision making in process of the protagonist. The first paragraph of the novel accustoms us slowly to these devices:

> The piano teacher, Erika Kohut, bursts like a whirlwind into the apartment she shares with her mother. Mama likes calling Erika her little whirlwind, for the child can be an absolute speed demon. She is trying to escape her mother. Her mother is old enough to be her grandmother. The baby was born after long and difficult years of marriage. Her father promptly left, passing the torch to his daughter. Erika entered, her father exited. (Ibid., 1988, 3)

Thus far, there is only a psychological problem posed, not yet a "mimesis of other minds," but when it comes we are given the problematic in these sentences, which is to what extent has our upbringing conditioned how we interact with others, especially at the level of primary passions. This problematic is stated in a sentence that has two meanings—one

---

7. See Paul Ricoeur, *Time and Narrative,* Volume 3, trans. Kathleen Blarney and David Pellauer (Chicago: University of Chicago Press, 1985), 163.
8. Paul Ricoeur, *Time and Narrative,* Volume 2, trans. Kathleen McLaughlin and David Pellauer (Chicago: University of Chicago Press, 1984), 90.
9. Elfriede Jelinek, *The Piano Teacher, A Novel,* trans. Joachim Neugroschel (New York: Grove Press, 1988).

historical, the other, arguably a here and now description: "Erika entered, her father exited." Although she has not seen him since her birth, we are cued that she has become the problematic focus of her mother's problems: "Her father promptly left, passing the torch to his daughter." The father is seen as one who makes quick decisions and carries them out with speedy behavior, energy filled "passing the torch" behavior.

The initial paragraph continues:

Eventually Erika learned how to move swiftly. She had to.

Here we have the first use of a "mimesis of other minds" demanded of the reader "She had to". This sentence is of the objective narrator, but it also can be reader as the internal voice of Erika reflecting on her immediate swift whirlwind movement described in the initial sentence of the novel. To cue mimesis, this art of description that occasions thought in the reader can be compared to the Living Theatre's shift from bodily anger, fear or suffering, to a movement toward and with the audience in a gestural or actual verbal challenge or invitation to respond. The initial paragraph continues:

Now she bursts into the apartment like a swarm of autumn leaves, hoping to get to her room without being seen. But her mother looms before her, confronts her. She puts Erika against the wall, under interrogation—inquisitor and executioner in one, unanimously recognized as Mother by the State and by the Family. She investigates: Why has Erika come home so late? Erika dismissed her last student three hours ago, after heaping him with scorn. You must think I won't find out where you've been, Erika. A child shown own up to her mother without being asked. But Mother never believes her because Erika tends to lie. Mother is waiting. She starts to count to three. (Ibid., 1988, 3–4)

We are given a deeper insight into the Mother's emotional violence which will be an analytical tool for the forthcoming passages that call upon our "mimesis of other minds."

The only emotion Erika is offered as she grows up into maturity is that of such violence. We are not given any details of her upbringing, the novel starting with this scene in her maturity as a piano teacher. She has no relationship with anyone but her students, other than the violence of her mother. Her teaching relationships are through elevating music, and the keyboard. At an early point in the novel, her student, a college-aged man, Walter Klemmer, will be invited subtly by her to have sex. "Klemmer" can be translated as one who pinches, who hurts. Klemmer is sexually inexperienced much younger than her, and, as we will see a masochist. Sensing in the intuitive manner of someone strongly influenced by a neurotic avenue of life, she opens herself up to immediate violence with Klemmer sexually, as that is the only way she can experience any emotion, given her upbringing. Yet, as we will see, this invitation will be in her mind a platform for a step toward her own power, autonomy and control over another—as she has been controlled by the power of her mother. Through excruciating scenes of mutual violence, she will come to a self-understanding of the perversion. Finally able to liberate herself through an agonized coming to know herself by the last pages, she will direct herself to the only source of self-liberation, where it all began, at home, where while the novel ends. As readers, we will have experienced through mimesis, how we individuals "make love," what are our constraints, what does leaving a partner after our love making mean. What have our relationships been? How do we leave? Though few readers will have had the experience we are shown with Erika, nonetheless, to read the novel is to find degrees of mimesis in our own experience, and efforts at self-knowledge. Indeed, perhaps coming to more knowledge of ourselves in the reading.

There is not sufficient room in this text for the details of this journey, which takes up almost two hundred pages in close description of the interactions between Klemmer and Erika. I will offer, nonetheless, several paragraphs that show her journey with Klemmer from beginning to end, with focus upon how sentences bring we as readers into self-understanding through mimesis, would we continue to read:

> Walter Klemmer pulls Erika out of the toilet stall. He yanks her. For openers, he presses a long kiss on her mouth; it was long overdue. He gnaws on her lips, his tongue plumbs her depths. After endlessly ruinous use, his tongue pulls back and then pronounces Erika's name several times. He puts a lot of work into this piece known as Erika. He reaches under her skirt, knowing that this means he is going places. He goes even farther, he feels that passion has permission. Passion has carte blanche. He burrows around in Erika's innards as if he wanted to take them out, prepare them in a new way. He reaches a limit and discovers that his hand can't get much farther. Now he pants as if he has run a great distance in order to reach this goal. He must at least offer this woman his exertion. He is unable to force his entire hand inside her, but maybe he can manage one or two fingers. No sooner said than done. Feeling his index finger slip in deeper than deep, he jubilantly transcends himself and bites Erika all over, promiscuously. He covers her with spit. His other hand holds her tight, but it doesn't need to, for the woman is staying put anyway. (Ibid., 1988, 176–77)

None of these sentences are of "erlebte Rede," as the narrator wants only bald description, not such intentional moments stemming from choice that are a threshold for reader "mimesis." That will come. But, first Erika's controlling responses in this sequence:

> Erika holds Walter Klemmer at arm's length. She pulls out his dick, which he has already slated for deployment. It only needs the finishing touch, for it is already prepared. Relived that Erika has taken over this difficult task, Klemmer tries to push his teacher down all the way. No Erika has to resist him with her entire weight so she can remain upright. She holds Klemmer's genital at arm's length while he fumbles about randomly in her vagina. She lets him know that if he doesn't stop, she'll leave. She softly repeats her threat several times, because her suddenly superior will has a hard time getting through to him and rutting fury. His mind seems fogbound with angry intentions. He hesitates. Wondering whether he's misunderstood something. Neither in the history of music nor anywhere else is the suitor simply barred from events. This woman has not a spark of submission. Erika starts kneading the red root between her fingers ... (Ibid., 1988, 178–79)

Two and one-half pages of this, then, Erika ceases, still in control. Then, "The student depicts the notorious painfulness of blue balls." This success for Erika leads to the next step of what Freud would call the dual expression of a symptom—furthering the illness, but yet being the praxis that if recognized can lead to self-healing. Erika writes a letter describing how she must be treated masochistically:

This paragraph says: Use a rubber hose—I'll show you how—to stuff the gag so tightly into my mouth that I can't stick out my tongue. Please use a blouse to increase my pleasure: tie up my face so skillfully and thoroughly that I can't get it off. And let me waste away in this torturous position for hours on end, so I can't do anything. I'll be stuck with myself and in myself. (Ibid., 1988, 218)

While the narration places this dialogue in a letter, sentence by sentence is what and how she says this to Klemmer. This is a form of "erlebte Rede," how we might speak to another in tone, if not in exact words. Her mother's torture of her is known in the blouse, and her limitation to being home always, except for her teaching. Klemmer fulfills these desires and more. He beats her, without emotion, coldly, in her room with the mother on the other side of the door, crying in rhythm with Erika. Again, there are many pages. This is not a salacious novel. Jelinek won the Noble Prize in Literature in 2004, and in the same year won the prestigious Franz Kafka prize. The ending of this novel, where Erika acts in such a way as to win insight into how to cure herself rather than engage again and again in torture and self-torture, refers to a well-known ending in Franz Kafka's *The Trial*. Erika, after the harshest beating by Klemmer, takes a knife to go find him, and as readers, we see that having experienced more than she could control, the ordinary "story" would call for such revenge as her killing him. She sees him at a distance, laughing among a group of fellow-students. But, here the Kafka enters, and her self-understanding:

Erika Kohut stands there, looking. She watches. It is broad daylight, and Erika watches. When the group has laughed its fill, it turns around toward the Engineering School. As the students move, they keep bursting into hardy guffaws. They interrupt themselves with their own laughter. (Ibid., 1988, 280)

The paragraph that follows is the penultimate paragraph, and contains the self-liberating insight, with the Kafka analog for such self-revelation:

Windows flash in the light. They do not open to this woman. They do not open to just anyone. There is no good person, although he is called for. Many would like to help, but do not. The woman twists here neck very far to the side and bares her teeth like a sick horse. No one puts a hand on her, no one takes anything from her. She feebly peers back over her shoulder. The knife should dig into her heart and twist around! (Ibid.)

Kafka 's penultimate paragraph in *The Trial* has him ponder his experience with the other-worldly visitations that had questioned his life path and actions throughout the novel. The novel is K.'s "trial" of how he perceives others and himself. It has been argued by critics that the entire novel is K's internal dialogue with his conscience and his search for actual love and caring with another, much as we see now with Erika Kohut. The final scene with the knife in *The Trial* covers many of the issues in the above penultimate paragraph of Jelinek:

Then one of them opened his frock coat and out of a sheath that hung from a belt girt round his waistcoat drew a long, thin, boublee-edged butcher's knife, held it up, and tested the cutting edges in the moonlight. ... K. now first perceived clearly that he was supposed to seize the knife himself, as it traveled from hand to hand above him, and plunge into his own breast. But he did not do so, he merely turned his head, which was still free, and gazed around him. He could not completely rise to the occasion, he could not relieve the officials of all their tasks; the responsibility for this last failure of his lay with him who had not left him the remnant of strength necessary for the deed ... With a flicker as of a light going up, the casements of a window there suddenly flew open; a human figure faint and insubstantial at that distance and that height leaned abruptly far forward and stretched both arms still farther. Who was it? A friend? A good man? Someone who sympathized? Someone who wanted to help? Was it one person only? Or was it mankind? Was help at hand? Were there arguments in his favor that had been overlooked? Of course there must be. Logic is doubtless unshakable, but it cannot withstand a man who wants to go on living. Where was the Judge whom he had never seen? Where was the high Court, to which he had never "penetrated?" He raised his hands and spread out all his fingers.

But the hands of one of the partners were already at K.'s throat, while the other thrust the knife deep into his heart and turned it there twice. With failing eyes K. could still see the two of them immediately before him, cheek learning against cheek, watching the final act. "Like a dog!" he said; it was as if the shame of it must outlive him.[10]

K. sees a light, and a window suddenly flies open. For Erika, the window does not fly open. It will not open for "just anyone." Only for he who was meant to enter "the law," although the waiter at the gate never knew until too late the entrance and the law was just for him—Kafka's famous parable. Yet, that was in an earlier chapter of *The Trial*. As the novel ends in the above scene it is the moment of insight for K. He does enter the law, just as we will see Erika enter it. There is no judge except for K. There is no judge for Erika except she herself. And, for the reader, he or she learns to be their sole judge in a "mimesis" of this moment. As the knife is thrust into his heart, it is he, the judge, his conscience that does it. The final lines in the German have him die "wie ein Hund."[11]

---

10. Franz Kafka, *The Trial*, trans. Willa and Edwin Muir (New York: Schocken Books, 1968), 228–29.
11. Jakob und Wilhelm Grimm, *Deutsches Wörterbuch*, Vierten Band, Zweiter Abtheilung [H.I.J.] (Leipzig, S. Hirzel, 1877), 1919. I offer the entire definitional passage because it allows one to see evidence of Kafka's adoption of some of its prose in an encoded manner that enables the reader to find this reference in Grimm:

> Hund, Hunde, m. centenarius, ein unterrichter item hant sie gewiesen, dasz die gerichtsherrn macht haben ihr geseig (gaichtes masz) zu geben klein und grosz, keins uszgenommen, und soll der schultheisz und hundt die von der herrn wegen geben. Weisth. 2, 30 (Saargegend, von 1421); und soll der hundte den (verbrecher) antworten an die buchenstaude. Das.; were sach, dasz die gerichtsherren wolten ihr jahrding oder ander gericht halten, so soll der schultheisz und der hundt ir jeglicher einen knecht haben, und dieselben knecht sollen da steen mit gewapneter handt. 32.

Kafka's last sentences in *Der Prozess* focus upon the Josef K's (Knecht's) hand, and then the hands of the court's assistants that appear to hold him and plunge the knife into his heart. With reference to the above, Josef K. is that knecht that is the armed, has a "gewapneter hand' so as to guard the administration of justice:

A synonym for judge in German is "Hund," not translated as so. Kafka wanted the initial understanding to be "dog," but wanted a more perceptive comprehension of "Hund" as well. His novel is one of encoded words, a centuries old tropic maneuver in German letters, employed by him. K.'s "shame" must outlive him—the final words. "Shame" can only outlive a misconceived self-image, as a correct one "shames" the past understanding. K. is not dead, rather a different person. Just as we will see Erika as a different person, subsequent to her plunging the knife in herself, with a flesh wound that is not mortal, but symbolic, as the direction of her path changes:

> Erika's back, where the zipper is partly open, is warmed. Her back is warmed by the ever more powerful sun. Erika walks and walks. Her back warms up in the sun. Blood oozes out of her. People look up from the shoulder to the face. Some turn around. Not all. Erika knows the direction she has to take. She heads home, gradually quickening her step. (Jelinek 1988, 280)

C. Group dynamics—The Focus Group

Charles E. Basch

Since the 1980s, the complexity of special problems in the practice of group dynamics has been studied in applied research by many small-group investigators. But, one or more of the key conceptual variables has continued to be the background guide of exploration. That is, the "four functional concerns" of Stimulation, Caring, Meaning and Executive Function (Lieberman, Yalom and Miles 1973, 447–48).

The "focus group" engages all four functional concerns. This is a form of group practice led by a group dynamics leader that began to be used in the fourth phase of the third Modern metaparadigm. The purpose was to elicit the opinion of "potential customers," of a product. Yet, as a form, it could be used for any inquiry into institutional practices, methodologies, or, in any human activities in a broader public that one wishes to have first-hand information about. Having arisen at the beginnings of group dynamics practice, this form took on early in its existence, which began in the late 1950s, a methodology that stressed input of an individual, and lesser interaction between those studied. One of the most thorough accounts of the "focus group" was written in 1987 by Charles E. Basch.[12]

Wo war der Richter, den er nie gesehen hatte? Wo war das hohe Gericht, bis zu dem er nie gekommen war? Er hob die Hände und spreizte all Finger. Aber an K.s Guirgel legten sich die Hände des einen Herrn, während der andere das Messer ihm tief ins Herz stiess und zweimal dort drehte. Mit brechenden Augen sah noch K., wie die Herren nahe vor seinem Gesicht, Wange an Wange aneinadnergelehnt, die Entscheidung beobachteten. "Wie ein Hund!" sagte er, es war als sollte die Scham ihn überleben. *(Der Prozess* in *Franz Kafka, Die Romane* (New York: Schocken Verlag, 1965, 444)

12. Charles E. Basch, "Focus Group Interview: An Underutilized Research Technique for Improving Theory and Practice in Health Education," *Health Education Quarterly* 14, no. 4 (Winter, 1987): 411–48.

Basch tells us that the focus group was discussed by E.S. Bogardus as early as 1926, and infrequently after that until the late 1950s (Basch 1987, 411, note 1). All of this began then in the fourth phase of the third Modern metaparadigm—with the idea of group interaction as its bases. Basch's use of the focus group, and its theoretical redirection, is an interesting study of the effect of phasal influence in the taking up of an idea from the collective to the more individual focus itself. Basch writes, for example, "In some respects, small groups as used by health educators are similar to processes involved in focus group interviews. The size of group is usually 6–12 (though this is flexible and productive groups can be conducted with fewer or more participants). The discussions usually last one to three hours; both require a comfortable, nonthreatening setting (physically and psychologically); the group leader or moderator requires skill in facilitating effective group functioning and is of key importance to the success full outcomes; and similar issues, such as concerns, experiences or problems, may be discussed" (ibid., 1987, 413–14).

Basch in his ongoing review of what has been done in his field will emphasize leadership insofar as the executive authority. There is not the minimum of intervention as in the T-Group of the 1950s through the mid-1960s, rather Basch gives the following responsibilities to the executive authority:

(1) If a group is unclear about its objectives, it will behave just as any individual does when his course of action is not clear to him.
(2) If clear-cut goals are lacking, the group cannot become cohesive, and the relations of each individual to the group becomes insecure, unproductive and perhaps meaningless.
(3) In a group where the goals are not clear, the high-status persons are those most likely to try to define goals for the group. In contrast, low-status members tend to withdraw from the group, to become unhappy and to devaluate their own effectiveness.
(4) Uncertainty as to the goal offers a kind of threat. Since the tensions aroused by this threat cannot be directed to the goal, they are released in other ways. Hostility and bad humor are often symptoms of goal uncertainty (ibid., 1987, 415–16).

Basch adds "Clarifying goals does not necessarily mean revealing the research hypothesis or questions under study. Clarifying goals does mean communicating to participants what you want to know from them" (ibid., 1987, 416). This understanding and practice negates the amplitude of "stimulation" allowed not only the leader, but interpersonally among the members, as well as the focus of "meaning" to that the leader wants to receive. In other words, thinking of the Living Theatre's pull-back to their conceptual interests in the 1970s, one will curtail what might be useful spontaneity insofar as what is learned from an more open-ended ability to respond. It is here that the next phase, the fourth phase of the collective, and more interpersonal, can augment the focus group.

A discussion outline is recommended by Basch:

Developing the discussion outline and questions to be used by the moderator requires careful thought and a considerable amount of effort. As in all questionnaire design, each item in the discussion outline should have a specific purpose. Items should relate to research aims, but certain items may also be included that are intended to facilitate the social and psycho-logical functioning of the group. Examples of facilitating items that can improve outcomes of groups are statements clarifying the goals of the study and assuring confidentiality, intro-ductory items that allow each participant to share something with the group, or strategically placed items that promote laughter and a relaxed climate. If threatening or sensitive topics are to be included, they should be preceded by nonthreatening items that allow the moder-ator and group members ample time to establish rapport with one another. As a general rule, items should proceed from general to specific. (Ibid., 1987, 417)

The discovery-innovative gain in the initial two phases of this fourth Modern metaparadigm, insofar as the democratic leveling voice of all, is absent from this research model. Here one sees the effects of tradition meeting the new models within a field. Certainly, the executive authority and stimulation lacks the flexibility that the Encounter group brought at its best to group interaction. Even, the previous fourth phase metaparadigmatic origins of the group encounter, is largely lost to the predefined horizon of outcomes expected. Basch reflects that limited horizon of outcomes as he counsels:

Reviewing literature related to the topic being investigated and consulting with content specialists may be useful for developing an instrument. (Ibid., 1987, 417)

Knowing one's field, certainly. Yet, consulting with content specialists may be too conser-vative in getting to know one's own experience with health issues, and what, after all, one must learn from the public about what effects their choices. The same criticism can be given Judith Malina. My sense of this entire study of Modernism is that a fourth phase has been deemed necessary by humanity. And that is because in formulating problems, and apply solutions, interpersonal, unsolicited information, is always vital when a broad approach is sought. The initial three phases of a metaparadigm enable the extremely cre-ative voice of individuals to sketch new approaches. But, as Charles Darwin said of the cultivation of special pigeon breeds, it is the "rock pigeon" who has been the same in sur-viving for millennia that must be considered in their collective, yet singular experiences. Group dynamics can integrate the new levels of self-reflection, but it must also provide the opportunity for those who self-reflect with a complex self-understanding to deepen the experiential knowledge of the group by taking it to a degree in directions not foreseen by those who lead the group.

Basch and the researchers he uses in his review of the value of focus groups on health education speak of mostly quantitative measures for determining relevance, derived from what a plethora of groups generate in their topics of discussion. Yet, the qualitative mate-rial that is reviewed for repetition by the many groups, thereby establishing a quantitative measure, does not offer a visible awareness of how they themselves limited what can be gleaned, by the strictness of their questionnaires and the group process (see, ibid., 1987, 419). When they speak of leader insight and use of what Basch, appropriately, calls "phe-nomenological" interventions, i.e., "reading facial expressions," however, in the service of "manipulating the defense mechanisms of group members" (ibid., 1987, 420). What is sought is still not full response in the acuity of the responder, rather: "homogeneity is important since intersubjectivity is unlikely to emerge if group members are dissimilar" (ibid.). One can explain this demand for "homogeneity" as the basis of a quantitative proof, but it is not the new metaparadigmatic call for singularity that indeed would provide significant new insights into how the public perceives personal health care.

Nonetheless, the language of the new level of being conscious, and the significance of methodologies that enable one to have a "mimesis of the other," is articulated, if not in the full awareness of why and how that "mimesis" has developed in the radical individuality of these initial phases in other disciplines. Basch writes under the "phenomenological" stress insofar as "methodology": "the moderator is actively involved in discourse in an attempt to affiliate with the group so he/she can truly see the world from their (our) perspective" (ibid.). Again, this is the traditionalist insofar as accepted conventions of thought and behavior.

Progress in human problem-solving has been indicated throughout the metaparadigms of Modernism. Nothing is ever finished or complete. New dimensions to what is known require new methodologies. These metaparadigmatic developments can be traced as far back in human history as records will allow, and promise increased knowledge and more thorough problem-solving as humanity moves forward.

CONCLUSION

The four evolving metaparadigms of Modernism have been shown in the four phases that recur in a spiral development from metaparadigm to metaparadigm, beginning about 1648 and continuing into the present. The rationale of the four phases of each metaparadigm is a common-sense finding by anyone who considers the stages of problem-solving in any endeavor. Since the evolutionary findings of Charles Darwin, in particular, the issue of how a species poses a problem, develops methods to address it, and how pragmatic attempts of instituting those methods, have been clear in any discipline. The normalization of methods that are successful are usually not marked by a discard of past discoveries, rather quite often either augment them, enable a more discerning comprehension of their possibilities within the present, or provide a significant corollary path of inquiry. The idea of individuation—the self-actualization of a person as they mature within an inherent set of ideas and capabilities, a scholastic discovery of the thirteen century by St. Thomas Aquinas, has been expanded in its dynamics in the thought of Leibniz, who added grammatical verbal evidence. Then, the thought of Friedrich Schleiermacher, a century later that formalized verbal grammatical evidence for individuation into the field of stylistics. Then, Wilhelm Dilthey and Edmund Husserl, at the beginning of the twentieth century, added the complex logic of an individual's manner of thought to the earlier stylistic expression. This taking of the problematic century after century is what I have referred to throughout this text as a "spiral" development of the same conceptual problem seen more finely, and its solution more finely developed.

Problem solving is the root of why distinct temporal periods are required in an individual's inquiry, and well as a collective group who share a discipline. As you the reader may reflect upon your own interests and their gradual development of breadth and depth, so it is how decades are often required simply to fashion the clarity of the problem through definition, and begin to articulate the ideation that may address what is viewed in a way that might solve it. You begin with certain principles, defining with clarity, often seeing how these principles have justified past approaches, but are capable of being applied in new ways. Then, developing a systematic approach with these principles, one can see system-building even in English and American pragmatism. Such as that in England's brilliant development of evolutionary theory from the empirical application of causal conceptions of the Darwins, grandfather and grandson, who used similar concepts systematically in differing orientations to the natural world they explained—the former in his use of intentionality and its consequences, and the latter in his use of how intentionality itself is determined. The application of these differing systematic directions for acquiring the evidence needed to either change ourselves and others, or, as with the grandson, to accommodate ourselves to the presence of others, has taken us as homo-sapiens to new levels of cultural progress.

The idea of a "metaparadigm"—that is, the shared conceptions and the methods they spawn among the arts and the sciences of a time—is also quite evident to ordinary reflection. How one has borrowed from several fields as one works upon one's disciplinary problems of a given time. As I have shown, investigators in the humanities, social sciences and the natural sciences have been inspired by the work of contemporaries in differing fields. Einstein's relativity theory stimulates both the literature of the early twentieth century in its stress of how a field forces differed from another, in the literature of Thomas Mann and his protagonists of singular perspectives, as well as the social sciences in Kurt Lewin's force field theory of interpersonal engagement.

Overarching as a premise in all metaparadigms of an epoch, such as Modernism, are certain axioms that are the larger aspects of human consciousness in species development. Modern's guiding axiom has been and still is "being aware of how one is aware" in greater and greater depths of epistemological knowledge of self and others. As we approach the end of not only this fourth metaparadigm, but arguably of our current epoch of Modernism, a new axiom has already emerged in the prescience of forward thinking minds, that is, we are now "aware of how we are aware in the constrained manner of one's particular cultural problematics in one's region of the planet". The structure of Judith Malina's *Paradise Now* that use the explanatory charts of the Chinese *I Ching*, the Indian Tantric teachings and the Hebrew Kabbalah (*Paradise Now* 1971, 1–3) are potential definitional exercises for a future multi-cultural heuristics. As the pre-Modern thought of a Montaigne whose reflections upon interpersonal conversation can be seen as a the forerunner of "being aware of how one is aware," Malina's multi-cultural study of stages of human conative praxis can be seen as a forerunner of a multi-culturalism, that may be the key idea that overarches the next epoch. Her inquiry into how one individuates in the philosophy of several cultures blend cultures into in-common interpretive tools into human conation with the outcome possible functional common practices derived from the history of these diverse cultures. If world peace is to be realized, common metaparadigms must become a global endeavor.

Many thinkers today, as well as more formally, historians of thought, are "metahistorians" without needing such a title or being aware that they are such. Mink's configurational comprehension of when ideas begin and later, and how the period of their dominance affects many differing fields of inquiry, can be seen in most historical accounts that include the guiding values and ideas of a time. Configuration comprehension means that the narrative includes the depth of certain attitudes and explanations throughout a culture of a time. The facts begin to be recognized as a pattern, the pattern is seen again and again.

Thomas Kuhn actually needed interpreters such as Margaret Masterman to formalize the idea of a "metaparadigm." Kuhn himself did not write much on the idea. While Kuhn articulates the concept of a metaparadigm several times in his *The Structure of Scientific Revolutions,* according to Margaret Masterman, who uses this term for the first time in describing Kuhn's theory, it required her stress on this fact for it to become a generalized concept for every age. I trust that this book will enable many practicing thinkers in differing fields, as well as historians and historiographers, to see work that they are already doing in the light of metahistory, and that this book will help them amplify their findings.

BIBLIOGRAPHY

Addison, Joseph. *The Spectator* in *The Complete Works of Joseph Addison*. Vol. V, ed. George Washington Greene. New York: G.P. Putnam, 1854.

Ankersmit, F. R. *Historical Representation*. Stanford, Calif.: Stanford University Press, 2001.

Ankersmit, Frank. *Meaning, Truth, and Reference in Historical Representation*. Ithaca, New York: Cornell University Press, 2012.

Ashley-Cooper, Anthony. The Third Earl of Shaftesbury. "Character." In *The Life, Unpublished Letters and Philosophical Regimen of Antony Earl of Shaftesbury*, edited by Benjamin Rand. London: Swan Sonnenschein, 1900.

Ashley-Cooper, Anthony. The Third Earl of Shaftesbury. *Characteristicks of Men, Manners, Opinions, Times*, Vol. One. Indianapolis: Liberty Fund, 2001.

Aristotle. *Eudemian Ethics*, ed. Brad Wood and Raphael Woolf. Cambridge: Cambridge University Press, 2013.

Aristotle. *Metaphysics*. The Basic Works of Aristotle, ed. Richard McKeon. New York: Random House, 1941.

Aristotle, *Nichomachean Ethics*. The Basic Works of Aristotle, ed. Richard McKeon. New York: Random House, 1941.

Aristotle. *Physics*. The Basic Works of Aristotle, ed. Richard McKeon. New York: Random House, 1941.

Babrius and Phaedrus, ed. Ben Edwin Perry. London: William Heinemann, 1965.

Bales, R.F. "Task and accumulation of experience as factors in the interaction of small groups." *Sociometry* 16 (1953): 239–52.

Balzac, Honoré De. *The Human Comedy*. https://en.wikipedia.org/wiki/La_Com%C3%A9die_humaine#Recurring_characters.

Balzac, Honoré De. *The Human Comedy*. Volume One. At the Sign of the Cat and Racket and Other Works. Trans. Clara Bell and R.S. Scott. Baltimore: Noumena Press, 2008.

Balzac, Honoré De. Pére Goriot. Trans. Wallace Fowlie. New York: Holt-Rinehart-Winston, 1960.

Bardi, Jason Socrates. *The Calculus Wars, Newton, Leibniz, and the Greatest Mathematical Clash of All Time*. New York: Thunder's Mouth Press, 2006.

Basch, Charles E. "Focus Group Interview: An Underutilized Research Technique for Improving Theory and Practice in Health Education." *Health Education Quarterly* 14, no. 4 (Winter 1987): 411–48.

Becker, Howard and Helmut Otto Dahlke. "Max Scheler's Sociology of Knowledge." In *Philosophy and Phenomenological Research* 2, no. 3 (March 1942): 310–22.

Bertalanffy, Ludwig von. *General System Theory*. Revised edition. New York: George Braziller, 1968.

Bloch, Eduard. https://en.wikipedia.org/wiki/Eduard_Bloch

Bloom, Harold. *The Western Canon*. New York: Harcourt, Brace, and Company, 1994.

Blum, Mark E. *Cognition and Temporality: The Genesis of Historical Thought in Perception and Reasoning*. New York: Peter Lang, 2019.

Blum, Mark E. *Continuity, Quantum, Continuum, and Dialectic: The Foundational Logics of Western Historical Thinking*. New York: Peter Lang, 2006.

Blum, Mark E. "Continuity and Discontinuity, Change and Duration: Hobbes' Riddle of the Theseus and the Diversity of Historical Logics." In *Theory and Research in Social Education* 24, no. 4 (Fall 1996): 360–90.

Blum, Mark E. *German and Austrian-German Historical Thought in the Modern Era.* New York: Lexington Books, 2020.

Boas, Franz. https://repository.si.edu/bitstream/handle/10088/13090/USNMP-11_709_1889. pdf?sequence=1&isAllowed=y

Bodin, Jean. *On Sovereignty,* ed. and trans. Julian H. Franklin. Cambridge: Cambridge University Press, 1992.

Boyer, Carl B. *The History of the Calculus and Its Conceptual Development.* New York: Dover, 1949.

Bradford, Leland, P., Jack R. Gibb and Kenneth D. Benne. *T-Group Theory and Laboratory Method, Innovation in Re-education.* New York: John Wiley & Sons, 1964.

Braunthal, Julius. *Viktor und Friedrich Adler, Zwei Generationen Arbeitrebewegung.* Vienna: Wienervolksbu chhandlung, 1965.

Brentano, Franz. *Psychology from an Empirical Standpoint.* Trans. Antos C. Rancurello, D.B. Terrell and Linda L. McAlister. Ed. Oskar Kraus and Linda L. McAlister. New York: Humanities Press, 1973.

Brentano, Franz. *The True and the Evident.* Trans. Roderick Chisholm, Ilse Politzer and Kurt R. Fischer. London: Routledge and Kegan Paul, 1971.

Carnap, Rudolf. *Introduction to Symbolic Logic and its Applications.* Trans. William H. Meyer and John Wilkinson. New York: Dover, 1958.

Cartwright, Dorwin and Alvin Zander, Eds., *Group Dynamics, Research and Theory,* 3rd edition. New York: Harper & Row, 1968.

Carr, David. *Time, Narrative, and History.* Bloomington: Indiana University Press, 1991.

Cassirer, Ernst. *Kant's Life and Thought.* Trans. James Haden. New Haven: Yale University, 1981.

Chomsky, Noam. *Aspects of a Theory of Syntax.* Cambridge, Mass.: M.I.T., 1965.

Darwin, Charles. *The Autobiography of Charles Darwin, 1809–1882.* New York: W.W. Norton, 1958.

Darwin, Charles. The Origin of the Species. In *Darwin, Texts, Backgrounds, Contemporary Opinion, Critical Essays,* ed. Philip Appleman. New York: Norton Critical Edition, 1979.

Darwin, Erasmus. *The Botanic Garden: A Poem in Two Parts,* Part I: the Economy of Vegetation. Miami, Fla: Hard Press, 2016.

Darwin, Erasmus. *Zoonomia; or, The Laws of Organic Life in Three Parts,* Complete in Two Volumes. Boston: D. Carlisle, 1803.

Defoe, Daniel. *The Fortunes and Misfortunes of the Famous Moll Flanders.* Oxford: Basil Blackwell, 1927.

Defoe, Daniel. *Robinson Crusoe.* New York: Bantam Books, 1981.

Diderot, Denis. *Thoughts on the Interpretation of Nature and Other Philosophical Works.* Trans. Lorna Sandler. Manchester: Clinamen Press, 1999.

Dilthey, Wilhelm. "Der Entwicklungs geschichtliche Pantheismus nach seinem geschichtlichen Zusammenhang mit den lllteren Pantheistischen Systemen."

Dilthey, Wilhelm. *Gesammelte Schriften,* Vol. II. Stuttgart: B.G. Teubner, 1957.

Dilthey, Wilhelm. *Introduction to the Human Sciences, An Attempt to Lay a Foundation for the Study of Society and History.* Trans. Ramon J. Betanzos. Detroit: Wane State University Press, 1979.

Dilthey, Wilhelm. *Leben Schleiermachers.* Two Volumes. Göttingen: Vandenhoeck & Ruprecht de Gruyter, 1966.

Dilthey, Wilhelm. *Poetry and Experience,* ed. Rudolf A. Makkreel and Frithjof Rodi. Princeton: Princeton University Press, 1985.

Dumain, Ralph. "The Autodidact Project: Quotes from Freud on Brentano (& Other Philosophers." http://www.autodidactproject.org/quote/freud_brentano.html

Dungey, Nicholas. "Shakespeare and Hobbes: Macbeth and the Fragility of the Political Order." SAGE Open April-June 2012: 1–18 http://sgo.sagepub.com.

Fielding, Henry. *Shamela.* New York: New American Library, 1980.

Franklin, Benjamin. *Benjamin Franklin Writings*. New York: The Library of America, 1987.

Freud, Sigmund. "The Antithetical Sense of Primal Words (1910)." In *Sigmund Freud, Collected Papers*. Trans. Joan Riviere. Volume 4. New York: Basic Books, 1959.

Freud, Sigmund. *Civilization and Its Discontents*. Trans. James Strachey. New York: W.W. Norton and Company, 1961.

Freud, Sigmund. *The Ego and the Id*. Trans. Joan Riviere. New York: W.W. Norton and Company, 1960.

Freud. Sigmund. "Hysterical Phantasies and Their Relation to Bisexuality." In *Sigmund Freud, Collected Papers*. Trans. Joan Riviere. Volume Two. New York: Basic Books, 1959.

Freud, Sigmund. "Notes Upon a Case of Obsessional Neurosis (1909)." In *Three Case Histories*, ed. Philip Rieff. New York: Collier Books, 1963.

Freud, Sigmund. "Some Elementary Lessons in Psycho-Analysis (1938)." In *Sigmund Freud, Collected Papers*. Vol. V, ed. James Strachey. New York: Basic Books, 1959.

Friedjung, Heinrich. *The Struggle for Supremacy in Germany, 1859–1866*. Trans. A.J.P. Taylor and W.L. McElwee. London: Macmillan & Co., 1935.

Gerard, Alexander. *An Essay on Taste*. New York: Garland, 1970.

Gianoutsos, Jamie. "Locke and Rousseau: Early Childhood Education." *The Pulse, the Undergraduate Journal of Baylor University* 4, no. 1 (2006).

Goethe, Johann Wolfgang von. "Elective Affinities," in *The Sufferings of Young Werther and Elective Affinities*. Trans. Elizabeth Mayer and Louise Bogan. New York: Continuum Publishing Company, 1991.

Goethe, Johann Wolfgang von. *The Sufferings of Young Werther*. Trans. Harry Steinhauer. New York: W.W. Norton, 1970.

Grimm, Jakob and Wilhelm. Deutsches Wörterbuch. Vierten Band, Zweiter Abtheilung [H.I.J.] Leipzig, S. Hirzel, 1877; 1919.

Grotius, Hugo. *The Rights of War and Peace, Including the Law of Nature and of Nations*. New York: Cosimo, 2007.

Gutmann, Hugo. https://en.wikipedia.org/wiki/Hugo_Gutmann

Haile, H.G. "'Octavia: Römische Geschichte,': Anton Ulrich's Use of the Episode." In *The Journal of English and Germanic Philology* 57, no. 4 (October 1958): 616ff.

Hanisch, Reinhold. http://marcuse.faculty.history.ucsb.edu/projects/hitler/sources/30s/394new rep/394NewRepHanischHitlersBuddy.htm

Hare, Paul. *Handbook of Small Group Research*. New York: The Free Press, 1962.

Hegel, Georg Wilhelm Friedrich. *Elements of the Philosophy of Right*, ed. Allen W. Wood. Trans. H.B. Nisbet. Cambridge: Cambridge University Press, 1991.

Hegel, Georg Wilhelm Friedrich. "Review, Proceedings of the Estates Assembly of the Kingdom of Württemberg, 1815–1816." In B. Bowman & A. Speight (Eds.), *Georg Wilhelm Friedrich Hegel: Heidelberg Writings: Journal Publications*. Cambridge: Cambridge University Press, 2009.

Heidegger, Martin. *An Introduction to Metaphysics*. Trans. Ralph Manheim. New York: Anchor Books, 1961.

Heidegger, Martin. "The Question Concerning Technology." *The Question Concerning Technology and Other Essays*. Trans. William Lovitt. New York: Harper Torchbooks, 1977.

Herder, Johann Gottfried. *Another Philosophy of History and Selected Political Writings*. Trans. Ionnis D. Evrigenis and Daniel Pellerin. Indianapolis: Hackett Publishing Company, 2004.

Herder, Johann Gottfried. "Auszug aus einem Briefwechsel über Ossian und die Lieder alter Völke." Von Deutscher Art und Kunst. Ed. Edna Purdie. Oxford: Oxford at the Clarendon Press, 1924.

Herder, Johann Gottfried. *Reflections on a Philosophy of the History of Mankind*. Chicago: University of Chicago Press.

Herder, Johann Gottfried. "Über die Reichsgeschichte, Ein historischer Spaziergang." Zur Philosphie der Geschichte, Eine Auswahl in Zwei Bänden. Berlin: Aufbau Verlag, 1952.

Herder, Johann Gottfried. Zur Philosophie der Geschichte, Eine Auswahl, in Zwei Bånden. Berlin: Aufbau Verlag, 1952.

Herodotus. *The Histories.* Trans. Aubrey de Sélincourt. Ed. A.R. Burn. Harmondsworth, Middlesex, England: Penguin, 1972.

Hobbes, Thomas. *The Correspondence of Thomas Hobbes.* Volume I: 1622–1659. Ed. Noel Malcolm. Oxford: Clarendon Press, 1997.

Hobbes, Thomas. *Dialogue, Behemoth, & Rhetoric.* Whithorn, England: Anodos Books.

Hobbes, Thomas. *Hobbes's Leviathan,* Reprinted from the Edition of 1651 (London: Oxford University Press, 1967), 8.

Hobbes, Thomas. https://nottinghamcityofliterature.com/blog/a-great-leap-in-the-dark-thomas-hobbes

Hobbes, Thomas. *Hobbes, On the Citizen,* ed. and trans. Richard Tuck and Michael Silverthorn. Cambridge: Cambridge University Press, 1997.

Hobbes, Thomas. *The Metaphysical System of Hobbes in twelve chapters from elements of philosophy concerning body,* ed. Mary Whiton Calkins. 2nd edition. LaSalle, Il.: Open Court, 1963.

Howes, Hetta Elizabeth. https://www.bl.uk/medieval-literature/articles/medieval-drama-and-the-mystery-plays

Husserl, Edmund. *The Crisis of the European Sciences and Transcendental Phenomenology: An Introduction to Phenomenological Philosophy.* Trans. David Carr. Evanston: Northwestern University Press, 1970.

Husserl, Edmund. *Logical Investigations.* Trans, J.N. Findlay. Vol. I. London: Routledge and Kegan Paul, 1970.

Ibsen, Henrik. *Peer Gynt and Brand.* Trans. Geoffrey Hill. England: Penguin Books, 2016.

Ibsen, Henrik. Pillars of Society (1877). In *Henrik Ibsen: The Complete Major Prose Plays.* Trans. Rolf Fjelde. New York: Plume Books, 1978.

Inwood, Stephen. *The Man Who Knew Too Much: The Strange and Inventive life of Robert Hooke 1635-1703.* London: Macmillan, 2002.

Jacob, Francois. *The Logic of Life: A History of Heredity.* Trans. Betty E. Spillman. New York: Pantheon Books, 1982.

Jaspers, Karl. *Nietzsche: An Introduction to the Understanding of His Philosophical Activity.* Trans. Charles K. Wallraff and Frederick J. Schmitz. Baltimore: Johns Hopkins University Press, 1997.

Jelinek, Elfriede. *The Piano Teacher: A Novel.* Trans. Joachim Neugroschel. New York: Grove Press, 1988.

Jencks, Charles. *The Story of Post-Modernism, Five Decades of the Ironic, Iconic and Critical in Architecture.* Chichester, West Sussex, UK: Wiley, 2011.

Kafka, Franz. Der Prozess. In Franz Kafka, Die Romane. New York: Schocken Verlag, 1965.

Kafka, Franz. *The Trial.* Trans. Willa and Edwin Muir. New York: Schocken Books, 1968.

Kant, Immanuel. "An Old Question Raised Again: Is the Human Race Constantly Progressing?", *Kant, On History.* Trans. Lewis White Beck. Indianapolis: Library of the Liberal Arts, 1963.

Kant, Immanuel. "Conjectural Beginnings of Human History," In *Kant, On History.* Trans. Lewis White Beck. Indianapolis: Library of the Liberal Arts, 1963.

Kant, Immanuel. "Idea for a Universal History From a Cosmopolitan Point of View." In *Kant, On History.* Trans. Lewis White Beck. Indianapolis: Library of the Liberal Arts, 1963.

Kant, Immanuel. Idee zu einer allgemeinen Geschichte in weltbûrgerlicher Absicht. In Immanuel Kants Werke, Schriften von 1783–1788. Hrsg. Artur Buchenau and Ernst Cassirer. Berlin: Verlegt bei Bruno Cassirer, 1922.

Kluback, William. *Wilhelm Dilthey's Philosophy of History.* New York: Columbia University, 1956.

Krieger, Leonard. *The Politics of Discretion: Pufendorf and the Acceptance of Natural Law.* Chicago: University of Chicago Press, 1965.

Kuhn, Thomas. *The Structure of Scientific Revolutions.* Chicago: University of Chicago Press, 1962.

Lamarck. https://en.wikipedia.org/wiki/Lamarckism#Transgenerational_epigenetic_inheritance

Leibniz, Gottfried Wilhelm. *New Essays on Human Understanding*, ed. by Peter Remnant and Jonathan Bennett. Cambridge: Cambridge University Press, 1996.

Leibniz, Gottfried Wilhelm. *Philosophical Letters and Papers*, ed. Leroy E. Loemker. 2nd edition. Dordrecht: D. Reidel Publishing Company, 1969.

Lessing, Gotthold Ephraim. *Hamburgische Dramaturgie*. Frankfurt am Main: Insel Verlag, 1986.

Lessing, Gotthold Ephraim. *Laocoön: An Essay on the Limits of Painting and Poetry*. Trans. Edward Allen McCormick. Baltimore: Johns Hopkins University Press, 1962.

Lessing, Gotthold Ephraim. Miss Sara Sampson, Ein bürgliches Trauerspiel in fünf Aufzügen. In *Lessings Werke in Fünf Bänden*. Berlin and Weimar: Aufbau Verlag, 1964.

Levine, Joseph. *Between the Ancients and the Moderns, Baroque Culture in Restoration England*. New Haven: Yale University Press, 1999.

Lewin, Kurt. https://en.wikipedia.org/wiki/Kurt_Lewin#Early_life_and_education

Lewin, Kurt. "Environmental Forces in Child Behavior and Development." In *A Dynamic Theory of Personality, Selected Papers*. Trans. Donald K. Adams and Karl E. Zener. New York: McGraw Hill Book Company, 1935.

Lewin, Kurt. *Principles of Topological Psychology*. Trans. Frtiz Heider and Grace M. Heider. New York: McGraw-Hill Book Company, 1936.

Lhotsky, Alphonse. *Österreichische Historiographie*. Vienna: Verlag für Geschichte und Politik, 1962.

Lieberman, Morton A., Irvin D. Yalom and Matthew B. Miles. *Encounter Groups: First Facts*. New York: Basic Books, 1973.

Locke, John. *Locke's Essay Concerning Human Understanding*. Books II and IV, ed. Mary Whiton Calkins. LaSalle, Ill.: Open Court, 1962.

Locke, *Two Treatises of Government*, ed. Peter Laslett. Cambridge: Cambridge University Press, 1988.

Lukács, György. https://en.wikipedia.org/wiki/Gy%C3%B6rgy_Luk%C3%A1cs#Communist_leader

Lucian. "Hermotimus, or the Rival Philosophies." *The Works of Lucian of Samasota*. Trans. H. Fowler and F.G. Fowler. Volume Two. Oxford: Clarendon, 1905.

Luther, Martin. https://blog.cph.org/read/everyday-faith/lutheranism/did-luther-really-say-here-i-stand

Malina, Judith. *The Enormous Despair: The Diary of Judith Malina, August 1968 to April 1969*. New York: Random House, 1972.

Malina, Judith. *The Diaries of Judith Malina, 1947–1957*. New York: Grove Press, 1984.

Malina, Judith. Interview. https://www.youtube.com/watch?v=VHYkIzgWv4Q

Malina, Judith and Julian Beck. *Paradise Now, Collective Creation of The Living Theatre*. New York: Vintage Book, 1971.

Malina, Judith. *Seven Meditations on Political Sado-Masochism*. https://www.youtube.com/watch?v=TZVtFiPNn34

Mann, Thomas. Death in Venice. In *Death in Venice and Seven Other Stories*. Trans. H.T. Lowe Porter. New York: Vintage Books, 1989.

Mann, Thomas. "Freud and the Future." In *Essays by Thomas Mann*. Trans. H. T. Lowe-Porter. New York: Vintage Books, 1957.

Marder, Michael. "Plant Intentionality and the Phenomenological Framework of Plant Intelligence." *Plant Signal Behavior* 7, no. 11 (November 2012): 1365–72.

Marx, Karl. Aus dem literarischen Nachlass von Karl Marx, Friedrich Engels und Ferdinand Lassalle. Hrsg. Franz Mehring. Vol. II. Gesammeltee Schriften von Karl Marx und Friedrich Engels von Juli 1844 bis November 1847. Stuttgart: J.H.W. Dietz Nachfolger, 1902.

Marx, Karl. Economic and Philosophic Manuscripts of 1844. In *Karl Marx and Friedrich Engels, Collected Works*. Volume 3. New York: International Publishers, 1976.

Marx, Karl. Ground Plan of the Critique of the Political Economy. In *The Portable Marx*, ed. Eugene Kamenka. Middlesex, England: Penguin Books, 1983.

Marx, Karl. Ökonomisch-philosophische Manuskripte. Frankfurt am Main: Suhrkamp Verlag, 2009.

Masterman, Margaret. "The Nature of a Paradigm, Criticism and the Growth of Knowledge." Proceedings of the International Colloquium in the Philosophy of Science. London, 1964. Vol. 4, ed. Imre Lakatos and Alan Musgrave. Cambridge: Cambridge University Press, 1970.

Montaigne, Michel de. "The Art of Conversation." *The Complete Essays*. Trans. M.A. Screech. London: Penguin Books, 1991.

Montesquieu. https://ouclf.iuscomp.org/montesquieu-in-england-his-notes-on-england-with-commentary-and-translation-commentary/

Montesquieu. *The Spirit of the Laws*. Trans. and Ed. Anne Cohler, Basia Miller and Harold Stone. Cambridge: Cambridge University Press, 1989.

Moreno, Jacob. https://davidmalocco.wordpress.com/2015/05/08/jacob-moreno-1899-1974-psychiatrist/

The New Scofield Study Bible. Authorized King James Version. New York: Oxford University Press, 1967.

Nietzsche, Friedrich. *The Advantage and Disadvantage of History for Life*. Trans. Peter Preuss. Indianapolis, Ind.: Hackett Publishing Company, 1980.

Ortega y Gasset, José. "Die Aufgabe unserer Zeit." Gesammelte Werke. Trans. Helene Weyl and Ulrich Weber. 4 vols. Stuttgart: Deutsche Verlags-Anstalt, 1950. *Oxford English Dictionary*. https://www.oed.com/view/Entry/63645?redirectedFrom=epoch#eid,

Pascal, Blaise. *Pascal's Pensées and The Provincial Letters*. New York: The Modern Library, 1941.

Pepper, Stephen C. *World Hypotheses, Prolegomena to Systematic Philosophy and a Complete Survey of Metaphysics*. Berkeley, Calif.: University of California Press, 1942; 1970.

Petrey, Sandy. "The Reality of Representation: Between Marx and Balzac." *Critical Inquiry* 14, no. 3, The Sociology of Literature (Spring, 1988): 448–68. Published by: The University of Chicago Press Stable URL: https://www.jstor.org/stable/1343698

Plato. Phaedrus in *The Collected Dialogues of Plato, Including the Letters*. Ed. Edith Hamilton and Huntington Cairns. Trans. R. Hackworth. Bollingen Series LXXI. Princeton: Princeton University Press, 1961.

Plato. Symposium in *The Collected Dialogues of Plato, Including the Letters*. Ed. Edith Hamilton and Huntington Cairns. Trans. Michael Joyce. Bollingen Series LXXI. Princeton: Princeton University Press, 1961.

Pufendorf, Samuel. *An Introduction to the History of the Principal Kingdoms and States of Europe*. Trans. Jodocus Crull (1695). Ed. Michael J. Seidler. Indianapolis: Liberty Fund, 2013.

Pufendorf, Samuel. *The Present State of Germany*. Trans. Edmund Bohum (1696). Edited with an Introduction by Michael J. Seidler. Indianapolis: Liberty Fund, 2007.

Ranke, Leopold von. Die grossen Mächte. http://www.gutenberg.org/files/39669/39669-h/39669-h.htm.

Reichenbach, Hans. *Elements of Symbolic Logic*. New York: Macmillan, 1948.

Richardson, Samuel. *Pamela; or, Virtue Rewarded*. New York: New American Library, 1980.

Ricoeur, Paul. *The Philosophy of Paul Ricoeur: An Anthology of His Work*. Ed. Charles E. Reagan and David Stewart. Boston: Beacon Press, 1978.

Ricoeur, Paul. *Time and Narrative*. Volume 2. Trans. Kathleen McLaughlin and David Pellauer. Chicago: University of Chicago Press, 1984.

Ricoeur, Paul. *Time and Narrative*. Volume 3. Trans. Kathleen Blarney and David Pellauer. Chicago: University of Chicago Press, 1985.

Rousseau, Jean-Jacques. *The Confessions of Jean-Jacques Rousseau*. Trans. J.M. Cohen. London: Penguin Books, 1953.

Rousseau, Jean-Jacques. *The Social Contract*. Trans. Maurice Cranston. Harmondsworth, Middlesex, England: Penguin Books, 1969.

Rousseau, Jean-Jacques. *Emile or Education*. Trans. Allan Bloom. Basic Books, 1979.

Sacher-Masoch, Leopold von. https://en.wikipedia.org/wiki/Legacy_of_Cain

Sacher-Masoch, Leopold von. Das Vermächtnis Kain. Erser Band. Zweiter unveränderter Abdruck. Stuttgart: Verlag der J.G. Cotta'schen Buchhandlung, 1870. Saint-Hilaire, Étienne Geoffroy. https://en.wikipedia.org/wiki/%C3%89tienne_Geoffroy_Saint- Hilaire#Geoffroy's_theory

Sartre, Jean-Paul. *Being and Nothingness*. Trans. Hazel E. Barnes. New York: Washington Square Press, 1966.

Sartre, Jean-Paul. *Critique of Dialectical Reason*. Trans. Alan Sheridan Smith. London: Verso, 1976.

Sartre, Jean-Paul. *The Family Idiot, Gustave Flaubert, 1821–1857*. 5 Volumes. Trans. Carol Cosman. Chicago: University of Chicago Press, 1981.

Sartre, Jean-Paul. No Exit in *No Exit and Three Other Plays*. New York: Random House, 1955.

Sartre, Jean-Paul. *Search for a Method*. Trans. Hazel E. Barnes. New York: Borzoi Books, 1967.

Scheler, Max. *Formalism in Ethics and Non-Formal Ethics of Values*. Trans. Manfred S. Frings and Roger L. Funk. Evanston, Ill.: Northwestern University Press, 1973.

Scheler, Max. Über Ressentiment und moralische Urteil. Leipzig: Wilhelm Engelmann, 1912.

Scheler, Max. Die Wissensformen und die Gesellschaft. Leipzig: Der Neue-Geist Verlag, 1926.

Schleiermacher, Friedrich. *Hermeneutics and Criticism and Other Writings*. Trans. Andrew W. Bowie. Cambridge: Cambridge University Press, 1998.

Schleiermacher, Friedrich. *Hermeneutics: The Handwritten Manuscripts*. Trans. James Duke and Jack Forstman. Atlanta, Georgia: Scholars Press, 1977.

Schnelle, Anna M. Die Staatsauffassung in Anton Ulrichs "Aramena" im Hinblick auf La Calprenédes "Cléopatre." Inaugural Dissertation zur Erlangung des Doktrgrades genehmigt von der Philosophischen Facultät der Friedrich-Wilhelm-Universität zu Berlin. Berlin: Verlag Rudolph Pfau, 1939.

Sebald, W. G. *Austerlitz*. Trans, Anthea Bell. London: Penguin Books, 2001.

Sengle, Friedrich. Das Genie und sein Fürst, Die Geschichte der Lebensgemeinschaft Goethes mit dem Herzog Carl August. Stuttgart: J.B. Metzlershe Verlagsbuchhandlung und Carl Ernst Poeschel Verlag, 1993.

Sokolowski, Robert. *Introduction to Phenomenology*. Cambridge: Cambridge University Press, 2000.

Spahr, Blake Lee. *Anton Ulrich and Aramena: The Genesis and Development of the Baroque Novel*. Berkeley and Los Angeles: University of California Press, 1966.

Spahr, Blake Lee. "Der Barockroman als Wirklichkeit und Illusion." Deutsche Romantheorien, Beiträge zu einer historischen Poetik des Romans in Deutschland. Hrsg. Reinhold Grimm. Frankfurt am Main: Athenäum Verlag, 1968.

Spencer, Herbert. *First Principles*. Second Edition. London: Williams and Norgate, 1867. The Stanford Encyclopedia of Philosophy, https://plato.stanford.edu/entries/kant/

St. John, Henry, Lord Bolingbroke. On the Idea of the Patriot King. In Bolingbroke's Letters on the Use and Study of History, etc. London: Ward, Lock & Co., N.D.

Stafeu, Frans A. *Linnaeus and the Linnaeans. The Spreading of their ideas in systematic botany, 1735–1789*. Utrecht: A. Oosthoek, 1971.

Sterne, Laurence. *Tristram Shandy*, ed. Howard Anderson. New York: W.W. Norton and Company, 1980.

Suarez, Francisco. Suarez on Individuation, Metaphysical Disputation V, Individual Unity and Its Principle. Trans. from the Latin with Introduction, Notes, Glossary, and Bibliography by Jorge J.E. Gracia. Milwaukee, Wisc.: Marquette University Press.

Sulloway, Frank J. *Freud, Biologist of the Mind: Beyond the Psychoanalytic Legend*. New York: Basic Books, 1979.

Ulrich, Anton. *Aramena*, ed. Dieter Merzbacher. Stuttgart: Anton Hiersemann, 2017.

Ulrich, Anton. *Die Römische Octavia*. Volume One, Part I. Ed. Rolf Tarot and Maria Munding. Stuttgart: Anton Hiersemann, 1993.

Wansink, Brian. "Changing Eating Habits on the Home Front: Lost Lessons from World War II." In *Journal of Public Policy & Marketing* 21, no. 1 (Spring 2002).

Weber, Max. https://en.wikipedia.org/wiki/Max_Weber

Weber, Max. "Zur Psychophysik der industriellen Arbeit." Part II. In Archiv für Sozialwissenschaft und Sozialpolitik, XXVIII Band, 1. Heft. January, 1909.

White, Hayden. *The Content of the Form, Narrative Discourse and Historical Representation*. Baltimore: Johns Hopkins University Press, 1987.

White, Hayden. *The Fiction of Narrative, Essays on History, Literature and Theory, 1957–2007*, ed. Robert Doran. Baltimore: Johns Hopkins Press, 2010.

White, Hayden. *Giambattista Vico: An International Symposium*, ed. Giorgio Tagliacozzo and Hayden White. Baltimore: The Johns Hopkins Press, 1969.

White, Hayden. *Metahistory: The Historical Imagination in Nineteenth-Century Europe*. Baltimore: Johns Hopkins University, 1973.

White, Hayden. *Tropics of Discourse: Essays in Cultural Criticism*. Baltimore: Johns Hopkins University Press, 1978.

Winckelmann, Johann Joachim. *Reflections on the Imitation of Greek Works in Painting and Sculpture*. Trans. Elfriede Heyer and Roger C. Norton. La Salle, Ill.: Open Court, 1987.

Wittgenstein, Ludwig. *Philosophical Investigations*. 3rd edition. Trans. G.E.M. Anscombe. New York: Macmillan, 1958.

Wolff, Friedrich Caspar. Theoria Generationis (1759), Part I, Vorrede, Erklärung des plans, Entwicklung der Pflaznen. Trans. Dr. Paul Samassa. Leipzig: Wilhelm Engelmann, 1896.

Wölfflin, Heinrich. *Principles of Art History: The Problem of the Development of Style in Later Art*. Trans. M.D. Hottinger. Mineola, N.Y.: Dover Publications, 1950.

Yolton, John W. "History and Meta-History." In *Philosophy and Phenomenological Research* 15, no. 4 (June 1955).

Young, Edward. https://en.wikipedia.org/wiki/Edward_Young

Young, Edward. *Conjectures on Original Composition*. Manchester: University of Manchester Press, N.D.

INDEX